INTRODUCTORY CHEMISTRY
Lecture Notes and Workbook

Second Edition

Kurt D. Donaldson
Jamie M. Scott

Edison Community College

Fort Myers, Florida

KENDALL/HUNT PUBLISHING COMPANY
4050 Westmark Drive   Dubuque, Iowa 52002

Copyright © 1992, 2001 by Kendall/Hunt Publishing Company

ISBN 978-0-7872-7696-6

All rights reserved. No part of this publication may be reproduced,
stored in a retrieval system, or transmitted, in any form or by any means,
electronic, mechanical, photocopying, recording, or otherwise,
without the prior written permission of the copyright owner.

Printed in the United States of America
4 5 6 7 8 9 10   08

CONTENTS

What is Chemistry...1
Atoms and Atomic Theory...25
Chemical Bonding 89
Chemical Calculations I...145
Chemical Reactions and Stoichiometry 189
Gases, Liquids, and Solids...225
Solutions and Solution Stoichiometry 257
Acids and Bases...297
Nuclear Chemistry 329
Chemical Nomenclature...353
Measurement and Significant Figures 367
Problem Solving Techniques 391

PREFACE

This book is the first in a series designed to improve the community college student's performance in introductory, general and organic chemistry. After years of teaching chemistry at the community college level, the authors have found certain key areas of difficulty for the student. These include poor note taking, not selecting the proper material for study, lack of attention to mathematical detail and insufficient review of the material. These books are written in such a way to help the student overcome these difficulties.

The content of this book is designed specifically for Edison Community College students in introductory chemistry. However, it is general and broad enough in its coverage that it will be adaptable to student needs at any school with a comparable course. The main text of each chapter is an outline of the lecture material. Key lecture topics are included with space provided for additional material presented by the instructor. Numerous worked out examples are given with space for additional examples. All examples involving calculations emphasize the proper set-up, mathematical mechanics and expression of the final answer. When a problem can be solved by more than one procedure the alternate procedures are also shown. At the end of each chapter a list of things to know is included. This includes as separate lists both important terms discussed in the chapter and important concepts and procedures. Following these lists is a set of review questions in fill-in-the-blank format. The last part of each chapter are review problems. Answers to the review problems are provided on the last page of each chapter.

The authors hope this book will help each student to perform better in their introductory chemistry class and help to develop proper problem solving abilities and study habits which will be valuable in future chemistry classes. The student should keep this completed book at the end of the course as it will be an excellent reference source for future courses and in the preparation of standardized tests such as the PCAT, MCAT, DCAT, etc. Good luck!

1 Unit One

What is Chemistry

Unit One

I. What is Chemistry?

A. **Chemistry** – the science of the composition, structure, properties, and reactions of matter, especially of atomic and molecular systems.

B. **Matter** – anything that has mass and occupies space (volume)

C. **Energy** – the capacity of matter to do work

$E = mc^2$ ← 186,000 mi/sec.

Mass vs. Weight

Mass – the amount of matter that an object possesses.

Weight – a measure of the effect of gravity on the object.

II. Matter — Properties and Changes

A. **Physical Properties vs. Chemical Properties**

Physical Properties – inherent physical characteristics of a substance that can be determined without altering its composition.

Chemical Properties – the ability of a substance to form new substances either by reaction with other substances or decomposition.

1. Examples of physical properties

   weight/mass, volume, color, physical state (solid, liquid, gas)

2. Examples of chemical properties

   $H_2O \xrightarrow{electricity} 2\ H_{gas} + 1\ O_{gas}$

B. **Intensive Properties vs. Extensive Properties**

Intensive Prop. – a property that is independent of the amount of material in a sample.

Extensive Prop. – a property that depends on the amount of material in a sample.

1. Examples of extensive properties

   heat (extrinsic) (mass, volume, pressure, etc.)

2. Examples of intensive properties

   temperature (intrinsic) (melt. temp / boiling temp / freezing temp,) density

C. **Physical Change vs. Chemical Change**

Physical Change – a change in which a substance changes from one physical state to another, but no substances with different compositions are formed.

Chemical Change – a change in which one or more new substances are formed.

INTRODUCTION TO CHEMISTRY

Unit One

III. Forms of Matter

A. One form of matter is the pure substance.

B. Characteristics of pure substances

C. There are two types of pure substances: elements and compounds

Elements vs. Compounds

Elements - a basic building block of matter that cannot be broken down into simpler substances by ordinary chemical ~~means~~ changes.

Compounds - a substance composed of 2 or more elements in fixed proportions, and can be decomposed into their constituent elements.

D. Another form of matter is the mixture.

E. Characteristics of mixtures

F. There are two types of mixtures: heterogeneous vs. homogeneous

Heterogeneous vs. Homogeneous

G. The breakdown of matter can be summarized as:

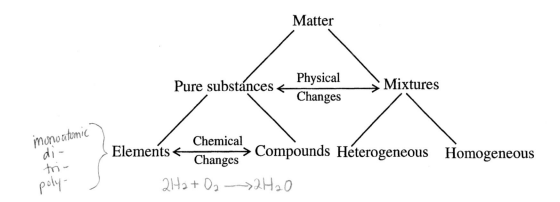

monoatomic
di-
tri-
poly-

$2H_2 + O_2 \rightarrow 2H_2O$

INTRODUCTION TO CHEMISTRY

Unit One

IV. More on Pure Substances and Mixtures

A. Most elements are naturally occurring, some are man-made.

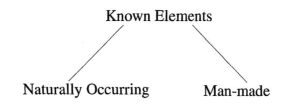

Known Elements
- Naturally Occurring
- Man-made

B. Abbreviations of elements

C. Compounds can be classified as covalent and ionic.

Covalent Compounds vs. Ionic Compounds

<u>Covalent compounds</u> - a chemical bond formed by the sharing of one or more electron pairs between two atoms.

<u>Ionic compounds</u> - a compound that is composed of ions.

D. Compounds are abbreviated by chemical formulas.

E. Chemical formulas

F. Another name for a <u>homogenous</u> mixture is a solution.

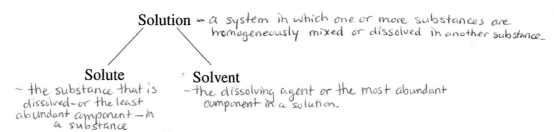

Solution — a system in which one or more substances are homogeneously mixed or dissolved in another substance

Solute — the substance that is dissolved — or the least abundant component — in a substance

Solvent — the dissolving agent or the most abundant component in a solution.

V. Energy — Forms and Changes

A. One way to classify energy is as either potential or kinetic.

Potential Energy vs. Kinetic Energy

<u>Potential Energy</u> - stored energy, or the energy of an object due to its relative position.

<u>Kinetic Energy</u> - the energy that matter possesses due to its motion.

INTRODUCTION TO CHEMISTRY

B. **Energy can also be classified by the type of work it does.**
   1. These include

   2. These forms of energy are interconvertible however the conversions are not 100% efficient
   3. Heat energy should not be confused with temperature

      Heat energy vs. Temperature

   4. Physical and chemical changes are accompanied by changes in heat energy
   5. Changes in heat energy are endothermic or exothermic

      Endothermic vs. Exothermic

   6. Changes in the amount of heat energy in matter will cause either a change in temperature, a change in state, or both

VI. States of Matter
A. **Matter can exist in three common states: gas, liquid, solid**
   1. Characteristics of gases

   2. Characteristics of liquids

   3. Characteristics of solids

VII. Changes in the State of Matter

A. Under the correct conditions matter will undergo a change in state

B. There are six possible changes in state

1. Melting

2. Freezing

3. Vaporization

4. Condensation

5. Sublimation

6. Deposition

C. For pure substances these changes occur at constant temperature

D. For mixtures these changes occur over a range of temperatures

E. Changes in state can be summarized as

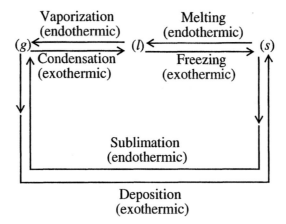

VIII. Conservation Laws

A. The Law of Conservation of Matter

B. The Law of Conservation of Energy

C. $E = mc^2$

D. The Law of Conservation of Matter and Energy

E. The Law of Conservation of Mass in Chemical Changes

Unit One

Things to Know

Definitions:

1. Chemistry - The science of the composition, structure, properties, and reactions of matter
2. Matter -
3. Energy
4. Physical property
5. Chemical property
6. Intensive property
7. Extensive property
8. Physical change
9. Chemical change
10. Pure substance
11. Element
12. Compound
13. Mixture
14. Heterogeneous mixture
15. Homogeneous mixture
16. Covalent compounds
17. Ionic compounds
18. Chemical formulas
19. Solution
20. Melting
21. Freezing
22. Vaporization
23. Condensation
24. Sublimation
25. Deposition
26. The Law of Conservation of Matter
27. The Law of Conservation of Energy
28. The Law of Conservation of Matter and Energy
29. Potential Energy
30. Kinetic Energy
31. Solid
32. Liquid
33. Gas
34. Heat energy
35. Temperature
36. Endothermic
37. Exothermic
38. Monoatomic element
39. Polyatomic element

INTRODUCTION TO CHEMISTRY

40. Molecule
41. Homonuclear molecule
42. Heteronuclear molecule
43. Electron sharing
44. Electron transfer
45. Cations
46. Anions
47. Empirical formulas
48. Solvent
49. Solute
50. Enthalpy
51. Changes in state
52. Changes in phase
53. Interparticle attraction
54. Free volume
55. Occupied volume
56. Mobile particles
57. Fluids
58. Condensed state
59. Crystalline structure
60. Specified distance
61. Specified angle

Things to Know

Concepts

1. Examples of physical properties
2. Examples of chemical properties
3. Examples of extensive properties
4. Examples of intensive properties
5. The two major types of matter
6. The two types of pure substances
7. The two types of mixtures
8. The number of known elements
9. The number of naturally occurring elements
10. How elements are abbreviated
11. How compounds are abbreviated
12. The classification of energies by work-type
13. The inter-conversion of energies
14. The effects of a change in heat energy of matter
15. What the various parts of $E = mc^2$ mean
16. How $E = mc^2$ changed the conservation laws

Review Questions

1. __Matter__ has mass and occupies space
2. __Energy__ is the ability to do work
3. An object's mass is __(?) constant (?)__ everywhere
4. An objects weight depends on the force of __gravity__
5. Mass, shape, color are examples of __physical__ properties
6. Combustibility is an example of a __chemical__ property
7. An example of an intensive property is __temperature__ (intrinsic)
8. An example of an extensive property is __heat__ (extrinsic)
9. Pure substances cannot be broken down by _____ means
10. The two types of pure substances are __elements__ and __compounds__
11. Which form of matter can be broken down by physical means _____
12. __Heterogeneous__ mixtures have different properties throughout
13. _____ cannot be broken down by physical or chemical means
14. _____ cannot be broken down by physical means but can be by chemical means
15. There are __113 or 114__ known elements
16. There are _____ naturally occurring elements
17. Elements are abbreviated by either _____ or _____ letter symbols
18. There are two types of compounds __covalent__ and __ionic__
19. Compounds are abbreviated by __chemical formulas.__
20. Chemical formulas give the _____ and _____ of atoms present
21. __Solution__ is another name for homogeneous mixture
22. Energy can be classified as __potential__ or __kinetic__
23. __Kinetic__ energy is the energy of motion
24. __Potential__ energy is the energy of position
25. When one form of energy converts into another the conversion is not _____ efficient
26. Energy "lost" in a conversion appears as _____ energy
27. _____ is a man-made scale measuring hotness or coldness
28. Changes in heat energy can be _____ or _____
29. __Exothermic__ changes release heat energy
30. __Endothermic__ changes absorb heat energy
31. Changes in the amount of heat energy in matter will cause either a change in _____, a change in _____ or both
32. The 3 states of matter are __solid__, __liquid__, __gas__

INTRODUCTION TO CHEMISTRY

33. __Solids__ have definite shape and definite volume
34. __Liquids__ have definite volume but assume the shape of their container
35. __Gases__ assume the shape of their container and completely fill it
36. A solid → liquid change in state is called __melting__
37. A liquid → solid change in state is called __freezing__
38. A liquid → gas change in state is called __vaporization__
39. A gas → liquid change in state is called __condensation__
40. A solid → gas change in state is called __sublimation__
41. A gas → solid change in state is called __deposition__
42. For a pure substance changes in state occur at __constant__ temperature
43. For mixtures changes in state occur over a __range__ of temperatures
44. Which 3 changes in state are endothermic __vaporization__, __melting__, __sublimation__
45. Which 3 changes in state are exothermic __condensation__, __freezing__, __deposition__
46. The total matter in the universe is constant is a statement of the _____
47. The Law of Conservation of Energy states that _____
48. Which formula shows the equivalence of matter and energy __$E=mc^2$__
49. The total matter and energy in the universe is constant is a statement of the _____
50. The Law of Conservation of Mass states that _____

2 Unit Two

Atoms and Atomic Theory

Unit Two

I. Elements

    A. Elements are pure substances which cannot be broken down by chemical change

    B. Known elements

    C. Naturally occurring elements

    D. Man-made elements

    E. Physical states of elements under natural conditions

    F. Elements are abbreviated by symbols

Names and Symbols of Some Common Elements

Element	Symbols	Element	Symbols	Element	Symbols
Carbon		Bromine		Sodium	
Hydrogen		Chlorine		Potassium	
Oxygen		Aluminum		Iron	
Nitrogen		Calcium		Copper	
Sulfur		Magnesium		Lead	
Phosphorus		Zinc		Silver	
Iodine		Silicon		Mercury	
Fluourine		Helium		Gold	

INTRODUCTION TO CHEMISTRY

II. Elements are composed of atoms

  A. Atom

  B. Origins of the atom concept

III. The Dalton Atomic Theory

  A. Postulates of the theory
  1.
  2.
  3.
  4.

  B. Is the theory still valid today

  C. The Law of Constant Composition

  D. The Law of Conservation of Mass

  E. Dalton and Berzelius developed the modern system of chemical symbolism

  F. Chemical formulas

  Examples: H_2O = 2 hydrogen atoms + 1 oxygen atom

  NH_3 = 1 nitrogen atom + 3 hydrogen atoms

  More examples:

Unit Two

G. **Empirical formula**

Examples: C_2H_4 becomes CH_2

$C_6H_{12}O_6$ becomes CH_2O

More examples:

IV. **Early atomic models**

A. **J. J. Thomson studied the properties of cathode rays in 1897**

B. **Schematic of cathode ray tube**

C. **Properties of cathode rays**

INTRODUCTION TO CHEMISTRY

D. Thomson proposes the first atomic model—The Plum Pudding Model

E. In 1909 Rutherford, a former student of Thomson, performed the Gold foil experiment

F. The discovery of radioactivity

G. Types of radioactivity

H. Schematic of the Gold foil experiment

I. What was expected to happen

J. What actually happened

K. Rutherford, in 1911, proposed the nuclear model of the atom
   1. The nuclear model

   2. The Rutherford nuclear model is unstable

V. Overview of Modern Atomic Theory

Particle	Discovery	Symbol	Charge	Mass
1. electron				
2. proton				
3. neutron				

A. Nuclear Symbolism $^{A}_{Z}X$

   1. Z

   2. A

B. Nearly all elements consist of more than one variety of atoms
   1. Isotopes

   2. The atomic mass unit

C. At the atomic level the whole does not equal the sum of the parts.
   Consider a C-12 atom:
   $$6\,p\,x$$
   $$6\,n\,x$$
   $$6\,e^{-}\,x$$

   The atomic mass of C-12 = 12.0000 amu
   Mass defect

D. Why is the nucleus stable?
   1. Nuclear binding energy

   2. Strong nuclear force

INTRODUCTION TO CHEMISTRY

VI. Background for the next atomic model

A. History of the nature of light
1. Huygens-Hooke
2. Newton
3. Young
4. Foucault
5. Maxwell
6. Hertz

B. Properties of waves
1. Amplitude
2. wavelength
3. frequency
4. speed
5. Light is an electromagnetic wave
6. The speed of light
7. The electromagnetic spectrum

C. Atomic Spectroscopy
1. Continuous spectrum

2. Absorption spectrum

3. Emission spectrum

D. **Blackbody radiation**
   1. Early attempts to explain it
   2. Planck, in 1900, did explain it
      a) Quanta
      b) $E = hf$

E. **The photoelectric effect**
   1. Early attempts to explain it

   2. Einstein, in 1905, did explain it

       Special relativity

       Brownian motion

F. **Is light a particle or is light a wave?**
   The Bohr principle of complementarity

VII. Neils Bohr, in 1913, proposed the Planetary model of the atom

A. **Goals of the model**
   1.
   2.
   3.
   4.

B. Postulates of the model
   1.
   2.
   3.

C. A general picture of the Bohr Planetary model

D. Shells, Orbits, Levels are synonymous

E. How many electrons can each orbit hold?

F. How did the Bohr model explain chemical periodicity?

Bohr model pictures for elements Z = 1 to Z = 20

$^{1}_{1}H$ $^{4}_{2}He$

$^{7}_{3}Li$ $^{9}_{4}Be$ $^{11}_{5}B$ $^{12}_{6}C$ $^{14}_{7}N$ $^{16}_{8}O$ $^{19}_{9}F$ $^{20}_{10}Ne$

$^{23}_{11}Na$ $^{24}_{12}Mg$ $^{27}_{13}Al$ $^{28}_{14}Si$ $^{31}_{15}P$ $^{32}_{16}S$ $^{35}_{17}Cl$ $^{40}_{18}Ar$

$^{39}_{19}K$ $^{40}_{20}Ca$

G. How did the Bohr model stabilize the atom?

H. How did the Bohr model include the "quanta" idea?

I. How did the Bohr model explain atomic spectroscopy?

J. The energy level diagram

K. What does it mean to have negative energy?

L. Moving up the energy level diagram requires energy

M. Moving down the energy level diagram releases energy

N. The Bohr model works mathematically only for H atoms

O. Fluorescence

P. Phosphorescence

Q. The wave nature of matter
   1. DeBroglie—1924

   2. Davisson, Germer, G. Thomson—1927

VIII. The modern theory of the atom—The quantum model—1925
   A. Dirac
   B. Heisenberg
   C. Schroedinger

   D. How are the electrons organized within the quantum atom?

   Bohr Model   vs   Quantum Model
   Quantum Numbers

Name	Symbol	Labels what	Determines what
1. Principal			
2. Orbital			
3. Magnetic			
4. Spin			

E. **Quantum numbers can be thought of as addresses of the electron**
   1. Each level consists of sublevels. Each sublevel consists of orbitals.

| Level | $n=1$ | $n=2$ | $n=3$ | $n=4$ |

Sublevels

Orbitals

   2. The Pauli Exclusion Principle

   3. How many electrons can each sublevel and level hold?

   4. The sublevels can be ordered by energy
   5. Notation

   6. Orbital

   7. Each type of orbital has a different shape.

      a) s orbitals

      b) p orbitals

      c) d orbitals

d) *f* orbitals

8. Probability plays a major role in the quantum theory

9. Heisenberg uncertainty principle

10. The Aufbau principle

11. Valence electrons

12. Building the elements in the quantum model

Element	Electron Configuration	Orbital Diagram	Number of Valence Electrons	Dot Formula

13. Hund's rule
14. Isoelectronic
15. Octet
16. The Periodic Law

IX. The Modern Periodic Table

Groups

Periods

A. Certain groups have special names

B. Certain periods have special names

C. Periodic Properties
1. Valence electrons
2. Metallic nature
   a) Properties of metals

   b) Properties of non-metals

c) Properties of metalloids

d) Any two elements can be compared as to metallic nature

Circle the most metallic element

    Na vs K   Mg vs Al   K vs Mg   Na vs Ca

3. Atomic size

a) The quantum atom is "fuzzy"

b) Down a group

c) Across a period

d) Any two elements can be compared as to atomic size

Circle the larger atom

    Na vs K   Mg vs Al   K vs Mg   Na vs Ca

4. Ionization energy

a) The 1st ionization energy

b) Down a group

c) Across a period

d) There are higher ionizations beyond the 1st

INTRODUCTION TO CHEMISTRY

e) The higher the ionization level the larger the energies required

5. Electron Affinity

   a) Down a group

   b) Across a period

Things to Know

Definitions
1. Elements
2. Physical states
3. Atom
4. Chemical formula
5. Empirical formula
6. Nucleus
7. Atomic number
8. Mass number
9. Isotopes
10. mass defect
11. nuclear binding energy
12. Amplitude
13. Wavelength
14. Frequency
15. Quantum
16. Shells, Orbits, Levels
17. Fluorescence
18. Phosphorescence
19. Valence e⁻ in the Bohr model
20. Valence e⁻ in the quantum model
21. Groups
22. Periods
23. Atomic size
24. 1st Ionization energy
25. Electron Affinity
26. Octet
27. Isoelectronic
28. Hadrons
29. Leptons
30. Quarks
31. Strong nuclear force

Scientists and Their Accomplishments

1. Democritus, Leucippus
2. Dalton
3. J.J. Thomson
4. Rutherford
5. Chadwick
6. Huygens-Hooke
7. Newton
8. Young
9. Foucault
10. Maxwell
11. Hertz
12. Bunsen, Kirchoff, Fraunhofer
13. Planck
14. Einstein
15. Bohr
16. DeBroglie
17. Dirac
18. Heisenberg
19. Schroedinger
20. Pauli
21. Mendeleev
22. Becquerel
23. Millikan

Concepts

1. Know names and symbols of the common elements
2. Know the postulates of the Dalton Atomic Theory
3. Know which parts have been modified
4. Know the Law of Constant Composition and how the Dalton theory explains it
5. Know the Law of Conservation of Mass and how the Dalton theory explains it
6. Be able to explain the meaning of chemical formula
7. Be able to convert a chemical formula into an empirical formula
8. How did Thomson discover the properties of cathode rays
9. Know the properties of cathode rays
10. Be able to describe Thomson's Plum Pudding model
11. Know the types of radioactivity
12. Describe the Gold foil experiment set-up
13. Know what was expected to happen
14. Know what actually did happen
15. Be able to describe the Rutherford Nuclear atom
16. Know why the nuclear atom is unstable
17. Know the types and properties of the sub-atomic particles (know the masses only approximately)
18. Be able to take a nuclear symbolism and determine # of p, # of n, # of e^- in the atom
19. Know why the nucleus is stable
20. Know the general properties of waves
21. Know the properties of light waves
22. Know the general make-up of the electromagnetic spectrum
23. Know the different types of atomic spectra
24. Know what blackbody radiation is and how it was explained
25. Know what the photoelectric effect is and how it was explained
26. Know the dual nature of light
27. Describe the Bohr Principle of Complementarity
28. Describe the Bohr Planetary Model
29. What were the goals of the Bohr model
30. Know how many electrons each orbit can hold
31. Be able to draw Bohr model pictures of a given atom
32. Know how the Bohr model stabilizes the atom
33. Know how the Bohr model included the quantum idea
34. Know how the Bohr model explained atomic spectroscopy results
35. Understand the energy level diagram representation of the atom
36. Understand what negative energy means
37. Know what happens when an electron moves up the energy level diagram
38. Know what happens when an electron moves down the energy level diagram
39. What was the major failure of the Bohr model
40. Be able to describe fluorescence in terms of energy level diagrams

41. Be able to describe phosphorescence in terms of energy level diagrams
42. Be able to describe the wave nature of matter
43. Know what quantum numbers in general mean
44. Know what each quantum number labels and determines
45. For $n = 1$ to $n = 4$ know the possible breakdown into sublevels and orbitals
46. Know how the Pauli Exclusion Principle relates to the filling of the orbitals
47. Know the probability nature of the orbital
48. Know the shapes of s and p orbitals
49. Know the different aspects of the Heisenberg uncertainty principle
50. Know how the Heisenberg uncertainty principle allows for violation of the conservation of matter and energy
51. Know what the Aufbau principle is
52. Know the orbital notation
53. Be able to give e⁻ configuration, orbital diagram, # of valence e⁻ and dot formula for $Z = 1$ to $Z = 36$
54. Know the arrangement of the Periodic Table
55. Know the names of the special groups
56. Know the names of the special periods
57. e able to get the # of valence e⁻ for main group elements
58. Be able to determine metallic nature by position in the Periodic Table
59. Know properties of metals, non-metals and metalloids
60. Be able to compare any two elements as to metallic nature
61. Know how atomic size varies down a group and across a period
62. Be able to compare any two main group elements as to size
63. Know how 1ˢᵗ ionization energy varies down a group and across a period
64. Be able to compare relative energies of different levels of ionization
65. Know how electron affinity varies down a group and across a period
66. Be able to derive long and short form e⁻ configurations from Periodic Table

Review Questions

1. Elements can't be broken down by physical or _____ change.
2. _____ and _____ are liquids at room temperature.
3. In one letter element symbols the letter is _____.
4. In two letter element symbols the first letter is _____ and the second letter is _____.
5. Elements are composed of _____.
6. An atom is the _____ part of an element retaining all of its properties.
7. _____ and _____ were early Greek philosophers who believed matter was discrete.
8. John Dalton developed the first _____ _____.
9. According to the Dalton theory atoms are _____.
10. According to the Dalton theory atoms of a given element are _____.
11. According to the Dalton theory during a chemical reaction atoms merely _____ themselves.
12. According to the Law of Constant Composition a given compound always contains the same _____ and _____ of atoms.
13. According to the Law of Conservation of Mass there is no _____ mass loss during a chemical reaction.
14. A _____ formula gives the number and types of atoms in a compound.
15. An _____ formula gives the type and lowest whole number ratio of atoms.
16. _____ studied the properties of cathode rays in 1897.
17. _____ discovered the electron in 1897.
18. Another name for the Thomson atomic model is the _____ model.
19. According to the Plum Pudding model the atom is a _____ mass of charge with the _____ embedded within.
20. _____ accidentally discovered radioactivity in 1896.
21. The three types of radioactivity are _____, _____, _____.
22. In 1909 _____ performed the _____ experiment.
23. Because of the unexpected results of the Gold Foil experiment Rutherford devised a new atomic model called the _____ model in _____.
24. The _____ is a very small positively charged region of the atom which contains most of the _____ of the atom.
25. In the nuclear model the electrons move _____ about the nucleus.
26. The Rutherford nuclear model is _____.
27. The _____ has charge = +1.
28. The _____ has charge = 0.
29. The _____ has charge = –1.
30. The proton was discovered by _____.
31. The neutron was discovered by _____.

INTRODUCTION TO CHEMISTRY

32. The number of protons in the nucleus of an atom is the _____.
33. The number of protons + neutrons in the nucleus is the _____.
34. Atoms with the same atomic number but different mass numbers are called _____.
35. Elements existing as isotopes is the _____ rather than the exception.
36. The _____ is 1/12th the mass of a C-12 atom.
37. The mass lost when protons and neutrons combined to form a nucleus is the _____.
38. The energy equivalent of the mass defect is the _____.
39. The force which holds the nucleus together is the _____.
40. The earliest particle theory of light was proposed by _____.
41. The earliest wave theory of light was proposed by _____.
42. _____ discovered the interference properties of light.
43. _____ first measured the speed of light.
44. _____ developed the mathematical theory of light waves.
45. _____ experimentally proved Maxwell's equations.
46. The height of a wave is called its _____.
47. The peak-to-peak distance in a wave is its _____.
48. The number of waves passing a given point in one second is the _____.
49. Light is a special type of wave called an _____ wave.
50. The highest energy EM waves are the _____.
51. X-rays are energetic enough to penetrate human soft tissue but not _____ tissue.
52. _____ rays are the cause of sunburns.
53. The colors of the visible spectrum from highest to lowest energy are _____, _____, _____, _____, _____, _____, _____.
54. Bright colored lines on a dark background is _____ spectrum.
55. Darklines on a continuous background is _____ spectrum.
56. Energy emitted by hot solid objects is called _____ radiation.
57. _____ explained blackbody radiation in 1900.
58. A particle of energy is called a _____.
59. _____ discovered the photoelectric effect.
60. _____ studied the photoelectric effect.
61. _____ explained the photoelectric effect.
62. According to the Bohr Principle of Complementarity light sometimes behaves like a _____, sometimes like a _____, but never _____ at the same time.
63. _____ proposed the Planetary model of the atom in 1913.
64. Bohr's goals for his model were: wanted a _____ atom, wanted to explain _____ periodicity, wanted to include the _____ idea, and wanted to explain _____.
65. According to Bohr electrons can exist in only certain _____ orbits.
66. The electron, in the Bohr model, can jump from one orbit to another in an _____.
67. The energy of an electron in the Bohr atom is _____.

INTRODUCTION TO CHEMISTRY

Unit Two

68. In the Bohr model 3 synonymous terms for the circular path of the electron are the _____, _____, _____.
69. The maximum number of electrons in the k shell is _____, in the L shell is _____, in the M shell is _____, in the N shell is _____.
70. The maximum number of electrons in any level, n, is given by _____.
71. Arranging the energy levels vertically is called an _____ diagram.
72. A negative energy of the electron means it is experiencing an _____ force.
73. Moving up the energy level diagram _____ energy.
74. The 3 main sources from which an electron can absorb energy are _____, _____, _____.
75. When all electrons are in the lowest possible energy level the atom is said to be in the _____ state.
76. When one or more electron is not in the lowest possible energy level the atom is said to be in the _____ state.
77. Moving down the energy level diagram _____ energy.
78. Energy released in an electron transition is always in the form of _____.
79. The average lifetime of an excited state is _____.
80. The main failure of the Bohr model is that it works _____ only for the atoms.
81. _____ is the absorption of high energy photons followed by the immediate release of lower energy photons.
82. _____ is the absorption of high energy photons followed by the delayed release of lower energy photons.
83. A _____ state is an excited state with an abnormally long lifetime.
84. _____ proposed the idea of matter waves.
85. According to DeBroglie the larger the mass the _____ the wavelength.
86. _____, _____, _____ experimentally confirmed matter waves.
87. _____, _____, _____ independently developed the quantum model.
88. _____ utilized the wave nature of the electron and was the most accepted approach.
89. Solving the Schroedinger _____ equation gives the _____ of the electron.
90. The wavefunction of an electron depends on certain numbers called _____ numbers.
91. Quantum numbers act as _____ for the electron in the atom.
92. The principal quantum number is symbolized by _____ and labels the _____.
93. The _____ quantum number is symbolized by l and labels the _____.
94. The magnetic quantum number is symbolized by _____ and labels the _____.
95. The principal quantum number determines the _____ and _____ of the orbital.
96. The orbital quantum number determines the _____ and _____ of the orbital.
97. The magnetic quantum number determines the _____ of the orbital.
98. Levels are composed of _____ which are composed of _____.

INTRODUCTION TO CHEMISTRY

Unit Two

99. The _____ principle states that the maximum number of electrons a given orbital can hold is two.
100. An orbital is a region around the nucleus with a _____ probability of finding an electron with the given set of quantum numbers.
101. All *s* sublevels can hold a maximum of _____ electrons.
102. All *p* sublevels can hold a maximum of _____ electrons.
103. All *d* sublevels can hold a maximum of _____ electrons.
104. All *f* sublevels can hold a maximum of _____ electrons.
105. All *s* orbitals are _____ in shape.
106. All *p* orbitals are _____ in shape and are oriented along the _____, _____, and _____ axes.
107. According to the _____ principle the more accurately you know the position the less accurately you can know the velocity of an electron.
108. _____ and _____ are related just as position and velocity by the Heisenberg Uncertainty principle.
109. According to the _____ principle when placing electrons into the sublevels they fill from lowest to highest energy.
110. In the quantum theory _____ electrons are those in _____ , and, _____ sublevels in the outer level of the atom.
111. A spin "up" electron is represented by _____.
112. A spin "down" electron is represented by _____.
113. According to _____ rule, when placing electrons into degenerate orbitals they remain _____ as long as possible.
114. Two or more species with the same electron configuration are said to be _____.
115. Atoms with 8 valence electrons are said to have an _____.
116. Vertical columns in the periodic table are called _____ or _____.
117. Horizontal rows in the periodic table are caled _____ or _____.
118. "A" group elements are called _____ group or _____ elements.
119. "B" group elements are called _____ elements.
120. Gp1A elements are called _____.
121. Gp2A elements are called _____.
122. _____ elements are called halogens.
123. _____ elements are called chalcogens.
124. Gp8A elements are called _____ or _____ gases.
125. For main group elements the number of valence electrons equals the _____.
126. _____ is the only metal which exists in the liquid state at room temperature.
127. _____ is the only non-metal which exists in the liquid state at room temperature.
128. Metals have _____ melting and boiling points.
129. Metals have _____, meaning they can be shined.

INTRODUCTION TO CHEMISTRY

Unit Two

130. Metals are _____, meaning they can be shaped.
131. Metals are _____, meaning they can be drawn into thin wires.
132. The diameter of an atom is difficult to measure because the quantum atom is _____.
133. Metallic nature _____ going down a group.
134. Metallic nature _____ going across a period.
135. Atomic size _____ going down a group.
136. Atomic size _____ going across a period.
137. The energy required to remove the outermost electron from an atom or ion in the gaseous state is the _____ energy.
138. The 1st ionization energy _____ going down a group.
139. The 1st ionization energy _____ going across a period.
140. The higher the ionizations the _____ the energies required.
141. The energy released or absorbed when a neutral gaseous atom gains an electron is the _____.
142. Going down a group electron affinity values _____.
143. Going across a period electron affinity value become more _____.

Review Problems

1. Interpret each formula in terms of numbers and types of atoms.
   a) NH_3
   b) CCl_4
   c) N_2O_4
   d) C_4H_{10}
   e) C_6H_6
   f) $C_{12}H_{22}O_{11}$
   g) OCl_2
   h) HBr
   i) N_2O_5

2. Convert each formula in question 1 to an empirical formula.

3. Determine the number of protons, neutrons, and electrons in each atom with the following nuclear symbol.
   a) $^{7}_{3}Li$
   b) $^{3}_{1}H$
   c) $^{15}_{7}N$
   d) $^{16}_{8}O$
   e) $^{40}_{18}Ar$
   f) $^{27}_{13}Al$
   g) $^{31}_{15}P$
   h) $^{235}_{92}U$
   i) $^{238}_{92}U$

4. Give Bohr model pictures for each and determine the number of valence electrons.
   a) $^{27}_{13}Al$
   b) $^{16}_{8}O$
   c) $^{15}_{7}N$
   d) $^{23}_{11}Na$
   e) $^{40}_{18}Ar$
   f) $^{20}_{10}Ne$

5. Give the breakdown into sublevels and orbitals for each level.
   a) $n = 1$
   b) $n = 2$
   c) $n = 3$
   d) $n = 4$
   e) $n = 5$

6. Give electron configuration, orbital diagram, number of valence electrons and dot formulas for:
   a) $_3Li$
   b) $_9F$
   c) $_{12}Mg$
   d) $_{24}Cr$
   e) $_{34}Se$

7. From the periodic chart determine the number of valence electrons for:
   a) $_{35}Br$
   b) $_{37}Rb$
   c) $_{51}Sb$
   d) $_{56}Ba$
   e) $_{54}Xe$

8. In each of the following pair circle the more metallic.
   a) K vs Ca
   b) Be vs Sr
   c) Rb vs Ca
   d) K vs Sr

9. In each of the following pairs circle the atom with the larger size.
   a) S vs Te
   b) N vs F
   c) P vs O
   d) N vs S

10. Write both the long form and short form e⁻ configuration for each:
    a) Pt
    b) Sb
    c) Ba
    d) At
    e) Ho
    f) Fm

Answers to Review Problems

1.
   a) 1 atom of N + 3 atoms of H
   b) 1 atom of C + 4 atoms of Cl
   c) 2 atoms of N + 4 atoms of O
   d) 4 atoms of C + 10 atoms of H
   e) 6 atoms of C + 6 atoms of H
   f) 12 atoms of C + 22 atoms of H + 11 atoms of O
   g) 1 atom of O + 2 atoms of Cl
   h) 1 atom of H + 1 atom of Br
   i) 2 atoms of N + 5 atoms of O

2.
   a) NH_3
   b) CCl_4
   c) NO_2
   d) C_2H_5
   e) CH
   f) $C_{12}H_{22}O_{11}$
   g) OCl_2
   h) HBr
   i) N_2O_5

3.
   a) #p = 3, #n = 4, #e⁻ = 3
   b) #p = 1, #n = 2, #e⁻ = 1
   c) #p = 7, #n = 8, #e⁻ = 7
   d) #p = 8, #n = 8, #e⁻ = 8
   e) #p = 18, #n = 22, #e⁻ = 18
   f) #p = 13, #n = 14, #e⁻ = 13
   g) #p = 15, #n = 16, #e⁻ = 15
   h) #p = 92, #n = 143, #e⁻ = 92
   i) #p = 92, #n = 146, #e⁻ = 92

4.
   a)
   b)
   c)
   d)
   e)
   f)

5. a)
```
        n = 1
          ↓
      1s sublevel
          ↓
     one 1s orbital
```

b)
```
             n = 2
           ↙      ↘
      2s sublevel   2p sublevel
          ↓              ↓
     one 2s orbital  three 2p orbitals
```

c)
```
               n = 3
           ↙      ↓      ↘
      3s sublevel  3p sublevel  3d sublevel
          ↓           ↓             ↓
     one 3s orbital  three 3p orbitals  five 3d orbitals
```

d)
```
                   n = 4
           ↙      ↙      ↘      ↘
      4s sublevel  4p sublevel  4d sublevel  4f sublevel
          ↓           ↓             ↓             ↓
     one 4s orbital  three 4p orbitals  five 4d orbitals  seven 4f orbitals
```

e)
```
                        n = 5
           ↙      ↙      ↓      ↘      ↘
      5s sublevel  5p sublevel  5d sublevel  5f  5g
          ↓           ↓             ↓        ↓    ↓
     one 5s     three 5p      five 5d   seven 5f  nine 5g
     orbital    orbitals      orbitals  orbitals  orbitals
```

6.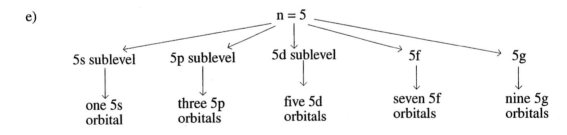

d) $1s^22s^22p^63s^23p^64s^13d^5$ ↑↓↑↓ (↑↓↑↓↑↓) ↑↓ (↑↓↑↓↑↓)↑(↑ ↑ ↑ ↑ ↑) ① Cr•

e) $1s^22s^22p^63s^23p^64s^23d^{10}4p^4$ ↑↓↑↓ (↑↓↑↓↑↓) ↑↓ (↑↓↑↓↑↓)(↑↓↑↓↑↓↑↓↑↓)(↑↓ ↑ ↑)

•S̈e:

7. a) 7   b) 1   c) 5
   d) 2   e) 8

8. a) K   c) Rb
   b) Sr   d) can't be determined

9. a) Te   c) P
   b) N   d) can't be determined

10. a) $_{78}$Pt   [Xe] $6s^24f^{14}5d^8$
    b) $_{31}$Sb   [Kr] $5s^24d^{10}5p^3$
    c) $_{56}$Ba   [Xe] $6s^2$
    d) $_{85}$At   [Xe] $6s^24f^{14}5d^{10}6p^5$
    e) $_{100}$Fm [Rn] $7s^25f^{11}6d^1$

INTRODUCTION TO CHEMISTRY

3 Unit Three

Chemical Bonding

Unit Three

I. Chemical Bonds

A. There are two types of chemical bonds: Ionic and Covalent

Ionic bonds vs Covalent bonds

B. The driving force for chemical bonding is the formation of an octet

C. The octet rule

II. Ionic Bonding

A. The attraction of oppositely charged ions produces the ionic bond

B. Ions

C. Ions come in two types: cations and anions

cations vs anions

D. Oxidation

E. Reduction

F. Cations are always smaller than their corresponding atoms

G. Anions are always larger than their corresponding atoms

H. Because of their low ionization energies metals lose e^- and form cations

I. Because of their large negative electron affinities non-metals gain e^- and form anions

J. Cations form with a charge making them isoelectronic with the preceding noble gas

K. Anions form with a charge making them isoelectronic with the suceeding

noble gas

L. **Properties of Ionic compounds**

M. **Naming of cations**

N. **Naming of anions**

O. **Given a metal and non-metal the formula of the resulting ionic compound can be determined by analyzing the electronic structure of each element and ion**

	Na	Cl	Mg	F
e⁻ configuration	$1s^22s^22p^63s^1$	$1s^22s^22p^63s^23p^5$	$1s^22s^22p^63s^2$	$1s^22s^22p^5$
# of valence e⁻	1	7	2	7
dot formula	Na•	$\ddot{\underset{\cdot\cdot}{Cl}}$	Mg:	$\ddot{\underset{\cdot\cdot}{F}}$
# of e⁻ lost or gained	lose 1e⁻	gain 1e⁻	lose 2e⁻	gain 1e⁻
dot formula of ion	[Na⁺]	$\left[\ddot{\underset{\cdot\cdot}{Cl}}\right]^-$	[Mg²⁺]	$\left[\ddot{\underset{\cdot\cdot}{F}}\right]^{1-}$
oxidation or reduction *rxn*	Na → Na⁺ + e⁻	Cl + e⁻ → Cl⁻	Mg → Mg²⁺ + 2e⁻	F + e⁻ → F⁻
dot formula of ionic compound		$[Na]^+\left[\ddot{\underset{\cdot\cdot}{Cl}}\right]^-$		$\left[\ddot{\underset{\cdot\cdot}{F}}\right]$ [Mg²⁺] $\left[\ddot{\underset{\cdot\cdot}{F}}\right]$
ionic formula		NaCl		MgF₂
name		sodium chloride		magnesium fluoride

Unit Three

	K	O	Al	S
e⁻ configuration				
# of valence e⁻				
dot formula				
# of e⁻ lost or gained				
dot formula of ion				
oxidation or reduction *rxn*				
dot formula of ionic compound				
ionic fomula				
name				

P. **When only an ionic formula is wanted, a short-cut procedure is to use the criss-cross rule**

Q. **Criss-cross rule**

R. **Transition metal ions**

INTRODUCTION TO CHEMISTRY

III. Properties of covalent compounds

A. Given a non-metal and a non-metal the formula of the resulting covalent compound can be determined by analyzing the electronic structure of each element.

	H	H	O	O	Cl	Cl
e⁻ configuration	$1s^1$	$1s^1$	$1s^2 2s^2 2p^4$	$1s^2 2s^2 2p^4$		
# of valence e⁻	1	1	6	6		
dot formula	H•	ˣH	:Ö•	ˣÖˣ (with x's around)		
# of e⁻ lost or gained	gain 1	gain 1	gain 2	gain 2		
dot formula of compound		H:H		:Ö:ˣÖˣ		
structural formula of compound		H – H		:Ö = Öˣ		
molecular formula		H_2		O_2		

INTRODUCTION TO CHEMISTRY

	F	F	Br	Br	N	N
e⁻ configuration						
# of valence e⁻						
dot formula						
# of e⁻ lost or gained						
dot formula of compound						
structural formula of compound						
molecular formula						

B. Covalent bonds can be classified as single, double or triple

single   double   triple

C. All of these compounds are classified as homonuclear diatomic molecules
1. homonuclear
2. diatomic

D. This type of analysis also works for heteronuclear diatomic molecules
heteronuclear

Unit Three

	H	Cl	H	Br
e⁻ configuration	$1s^1$	$1s^2 2s^2 2p^6 3s^2 3p^5$		
# of valence e⁻	1	7		
dot formula	H•	ˣ:Cl:ˣ (with x's around)		
# of e⁻ lost or gained	gain 1	gain 1		
dot formula of compound		H :Cl:		
structural formula of compound		H—Cl:		
molecular formula		HCl		

E. **Shared pairs of electrons are not always shared equally**

F. **Electronegativity**

G. **Electronegativity is a periodic property**
   1. Down a group

   2. Across a period

H. **When the two bonded atoms have the same electronegativity the sharing is equal**
   Examples
   2.1 2.1
   H : H
   equal attraction creates
   equal sharing

INTRODUCTION TO CHEMISTRY

I. **When the two bonded atoms have different electronegativities the sharing is unequal**
Examples
2.1   3.0

H ⦂Cl

unequal attraction creates
unequal sharing

J. **Unequal sharing creates partial charges within the molecule**
Examples

$$H^{\delta^+} : Cl^{\delta^-}$$

δ^+ read "delta plus" = partial positive charge

δ^- read "delta minus" = partial negative charge

K. **Partial charges within a molecule create a dipole**
Dipole

L. **A dipole within a molecule is represented by an arrow in the structural formula pointing from positive to negative.**
Examples

$$H^{\delta^+} \to Cl^{\delta^-}$$

M. **Covalent bonds can be classified as non-polar or polar**

non-polar vs polar

N. The sharing of electrons occurs by the overlap of atomic orbitals.

O. There are two types of overlap: σ (sigma) and π (pi)

σ vs π

P. For our purposes, only orbitals with an unpaired e⁻ can be involved in orbital overlap.

Q. Orbital overlap of some diatomic molecules

| H_2 | Cl_2 | HCl | O_2 | N_2 |

H—H   Cl—Cl
σ(1s – 1s)   σ(3p – 3p)

IV. The structural formula of molecules with more than two atoms cannot be so easily determined by the above procedure.

A. Two slightly different but equivalent methods can be used for these molecules called polyatomic molecules
   1. N—A = S method

2. Skeletal structure method

B. Lewis structures of some common polyatomic molecules
   1. CH_4
   2. NH_3

Unit Three

3. H_2O

4. $BeCl_2$

5. BF_3

6. SO_2

Some combinations of atoms have a net charge and are called polyatomic ions. The Lewis structure of a polyatomic ion can be determined just as for a polyatomic molecule except the charge of the ion will change the number of valence e⁻ available. Lewis structures of some common polyatomic ions

1. NO_3^{1-}

2. ClO_3^{1-}

3. CO_3^{2-}

4. SO_4^{2-}

Unit Three

V. The shape of a polyatomic molecule or ion can be determined by a method known as VSEPR

    A. VSEPR stands for

    B. The VSEPR method

Once the correct shape of a molecule is determined each atom can be assigned a partial positive or negative charge, where applicable, and each bond represented as a dipole, again where applicable.

The overall polarity of a molecule depends on the cancellation or addition of the dipoles associated with individual bonds.

    C. There are two types of geometries of molecules: electronic and molecular

        electronic geometry vs molecular geometry

    D. There are two types of e⁻ pairs around the central atom: bonding (b or bp) and non-bonding (nb or nbp)

    E. There are three types of repulsions possible between these:
        b — b < b — nb < nb — nb

Unit Three

F. Examples of using VSEPR method

	1. CH₄	2. NH₃	3. H₂O	4. BeCl₂
Lewis structure				
VSEPR class				
Electronic geometry				
Molecular geometry				
Bond angles				
Polarity				
Correct shape showing partial charges and dipoles				

	5. BF₃	6. SO₂	7. NO₃¹⁻	8. ClO₃¹⁻
Lewis structure				
VSEPR class				
Electronic geometry				
Molecular geometry				
Bond angles				
Polarity				
Correct shape showing partial charges and dipoles				

Unit Three

Summary of Molecular properties based on VSEPR

VSEPR class	Electronic geometry	Molecular geometry	Bond angles	Polarity	Hybridization of central atom
1. AB_2					
2. AB_3					
3. AB_2E					
4. AB_4					
5. AB_3E					
6. AB_2E_2					

Additional Topics

1. Representing non-polar and polar molecules

2. Hybrid orbitals

INTRODUCTION TO CHEMISTRY

Things to Know

Definitions
1. ionic bond
2. covalent bond
3. octet rule
4. ion
5. cation
6. anion
7. oxidation
8. reduction
9. single bond
10. double bond
11. triple bond
12. homonuclear diatomic
13. heteronuclear diatomic
14. electronegativity
15. non-polar bond
16. polar bond
17. partial charge
18. dipole
19. σ overlap
20. π overlap
21. polyatomic molecule
22. polyatomic ion
23. bond angle
24. electronic geometry
25. molecular geometry
26. bonding e⁻ pair
27. non-bonding e⁻ pair
28. covalency
29. dipole moment

Concepts

1. Know how cation size compares to parent atom size
2. Know how anion size compares to parent atom size
3. Know why metals tend to lose e^-
4. Know why non-metals tend to gain e^-
5. Know what stable cation charges are isoelectronic to
6. Know what stable anion charges are isoelectronic to
7. Know the properties of ionic compounds
8. Be able to name cations of main group metals
9. Be able to name cations of transition metals
10. Be able to name monoatomic anions
11. Given a metal and non-metal be able to deduce ionic formula by electronic structure analysis
12. Given a metal and non-metal be able to deduce ionic formula by criss-cross rule
13. Given two identical non-metals be able to deduce structural and molecular formula by electronic structure analysis
14. Know properties of covalent compounds
15. Be able to identify and distinguish single, double, and triple bonds
16. Given two different non-metals be able to deduce structural and molecular formula by electronic structure analysis
17. Know the periodic variation of electronegativity
18. Know how atom electronegativities, partial charges and dipoles are related and symbolized
19. Know which type of orbitals are involved in σ overlap
20. Know which type of orbitals are involved in π overlap
21. Be able to determine Lewis structural formula of polyatomic molecules by either method
22. Be able to determine Lewis structural formula of polyatomic ions by either method
23. Know what VSEPR stands for
24. Given a Lewis structural formula be able to classify within the VSEPR method — $A B_N E_M$
25. From a VSEPR class be able to predict electronic and molecular geometries, bond angles and polarity
26. From VSEPR class be able to draw a molecule or ion with correct shape, partial charges and dipoles
27. Be able to analyze transition metal ions as to charges and e^- configuration
28. Know the importance of a molecule being non-polar or polar and how we represent such molecules
29. Know the properties of hybrid orbitals and be able to predict the hybridization scheme of the central atom from its VSEPR classification

Review Questions

1. The driving force for a chemical reaction is the attainment of an _____.
2. The transfer of electrons produce an _____ bond.
3. The sharing of electrons produce a _____ bond.
4. When an atom loses or gains one or more electrons it produces an _____.
5. A positively charged ion is called a _____.
6. A negatively charged ion is called an _____.
7. The loss of one or more electrons during a chemical reaction is called _____.
8. The gain of one or more electrons during a chemical reaction is called _____.
9. Cations are always _____ than their corresponding atom.
10. Anions are always _____ than their corresponding atom.
11. Because of their _____ ionization energies metals _____ electrons and form _____.
12. Because of their large _____ electron affinities non-metals _____ electrons and form _____.
13. Cations have charges making them _____ with the _____ noble gas.
14. Anions have charges making them _____ with the _____ noble gas.
15. Ionic compounds consist of _____ and _____.
16. Ionic compounds must be electrically _____.
17. In the solid state ionic compounds exist in a _____.
18. Ionic compounds have _____ melting and boiling points.
19. In the solid state ionic compounds are _____ conductors of heat and electricity.
20. When melted ionic compounds are _____ conductors of heat and electricity.
21. Main group metals form ionic compounds which are usually _____.
22. Transition metals form ionic compounds which are usually _____.
23. Main group metal ions are named using the _____ name.
24. Transition metal ions are named using the _____ name followed by the charge in numerals in _____.
25. Anions are named using the _____ of the element name and the suffix _____.
26. Covalent compounds consist of _____ and _____ atoms.
27. Covalent compounds must be electrically _____.
28. Covalent compounds exist as _____.
29. Covalent compounds have _____ melting and boiling points.
30. Only _____ electrons can be shared with other atoms.
31. _____ is an exception to the octet rule and forms compounds with only 2 valence electrons.
32. One pair of shared electrons produces a _____ covalent bond.

Unit Three

33. Two pairs of shared electrons produces a _____ covalent bond.
34. Three pairs of shared electrons produces a _____ covalent bond.
35. H_2 and Cl_2 are examples of _____ _____ molecules.
36. HCl is an example _____ _____ molecules.
37. _____ measures the attraction of an atom for shared electrons.
38. The electronegativity scale runs from _____ to _____.
39. Going down a group electronegativity _____.
40. Going across a period electronegativity _____.
41. Equal attraction means _____ sharing.
42. Equal sharing of electrons produces a _____ covalent bond.
43. Unequal attraction produces _____ sharing.
44. Unequal sharing produces a _____ covalent bond.
45. Unequal sharing creates _____ charges within the molecule.
46. The atom with the higher electronegativity in a polar bond has the partial _____ charge.
47. The atom with the lower electronegativity in a polar bond has the partial _____ charge.
48. Partial charges within a molecule create a _____.
49. A dipole is the separation of _____ and _____ charges.
50. A dipole within a molecule is represented by an _____ pointing from _____ to _____.
51. The sharing of electrons occurs by the _____ of atomic orbitals.
52. The two types of orbital overlap are _____ and _____.
53. When at least one s orbital is involved the orbital overlap is _____.
54. Two p orbitals overlapping head-to-head produce _____ type of overlap.
55. Two p orbitals overlapping side-to-side produce _____ type of overlap.
56. In the VSEPR theory the electron pairs around the central atom arrange themselves into the geometry which _____ their mutual repulsion.
57. The arrangement of the electron pairs around the central atom creates the _____ geometry.
58. The arrangement of the atoms creates the _____ geometry.
59. An AB_2 molecule has electronic geometry of _____ and molecular geometry of _____.
60. An AB_3 molecule has electronic geometry of _____ and molecular geometry of _____.
61. An AB_2E molecule has electronic geometry of _____ and molecular geometry of _____.

INTRODUCTION TO CHEMISTRY

Unit Three

62. An AB_4 molecule has electronic geometry of _____ and molecular geometry of _____.
63. An AB_3E molecule has electronic geometry of _____ and molecular geometry of _____.
64. An AB_2E_2 molecule has electronic geometry of _____ and molecular geometry of _____.
65. An AB_2 molecule has bond angles _____ and is _____.
66. An AB_3 molecule has bond angles _____ and is _____.
67. An AB_2E molecule has bond angles _____ and is _____.
68. An AB_4 molecule has bond angles _____ and is _____.
69. An AB_3E molecule has bond angles _____ and is _____.
70. An AB_2E_2 molecule has bond angles _____ and is _____.
71. The bond angles in AB_2E and AB_3E are depressed because the repulsion between non-bonding and bonding electron pairs is _____ the repulsion between bonding and bonding electron pairs.
72. The bond angles in AB_2E_2 are depressed even more because the repulsion between non-bonding and non-bonding electron pairs is _____ the repulsion between non-bonding and bonding electron pairs.
73. _____ is an exception to the octet rule and will form compounds with only 6 valence electrons.
74. _____ is an exception to the octet rule and will form compounds with only 4 valence electrons.

Review Problems

1. Circle the largest in size in each pair.
   a) K vs K^+
   b) Ca vs Ca^{2+}
   c) O vs O^{2-}
   d) Cl vs Cl^{1-}

2. Which noble gas is each ion isoelectronic to
   a) Na^{1+}
   b) Ca^{2+}
   c) Al^{3+}
   d) Br^{1-}
   e) Se^{2-}
   f) P^{3-}

3. For each metal and non-metal pair give 1) electron configuration 2) number of valence electrons 3) Lewis dot formula 4) number of electrons lost or gained 5) dot formula of ion 6) oxidation or reduction reaction 7) dot formula of ionic compound 8) ionic formula
   a) Na and F
   b) Mg and S
   c) Ca and N
   d) Al and Br

4. Use the criss cross rule to determine the ionic formula for the combinations.
   a) Li and O
   b) Ba and P
   c) Al and S
   d) K and N
   e) Ca and S
   f) Al and P

5. For each non-metal and non-metal combination determine 1) electron configuration 2) number of valence electrons 3) Lewis dot formula 4) number of electrons lost or gained 5) Lewis dot formula of the compound 6) Lewis structural formula of the compound 7) molecular formula
   a) I and I
   b) O and O
   c) N and N
   d) Cl and Cl

6. For each of the following molecules determine the type of orbital overlap and sketch it.
   a) HBr
   b) O_2
   c) N_2
   d) Br_2
   e) HF

7. For each of the following molecules determine its Lewis dot and structural formulas.
   a) PH_3
   b) CCl_4
   c) BeF_2
   d) H_2Se
   e) BCl_3
   f) SO_3

8. For each of the following polyatomic ions determine its Lewis dot and structural formulas.
   a) BrO_3^{1-}
   b) NO_2^{1-}
   c) SO_3^{2-}
   d) PO_4^{3-}
   e) NH_4^+

9. For each of the following molecules or ions determine 1) VSEPR class 2) electronic geometry 3) molecular geometry 4) bond angles 5) polarity 6) Draw with correct shape showing partial charges and dipoles.

a) CCl_4

b) PI_3

c) H_2Se

d) BeF_2

e) BCl_3

f) SO_3

g) NO_2^{1-}

h) BrO_3^{1-}

i) SO_3^{2-}

j) PO_4^{3-}

k) NH_4^+

Answers to Review Problems

1. a) K   c) O^{2-}
   b) Ca d) Cl^{1-}

2. a) Ne c) Ne e) Kr
   b) Ar d) Kr f) Ar

3. a) Na   F
   1) $1s^22s^22p^63s^1$ $1s^22s^22p^5$

   2) ①   ⑦

   3) Na•   ⨯F⨯ (dot structure)

   4) lose 1   gain 1

   5) $[Na^{1+}]$ $\left[\text{⨯F⨯} \right]^{1-}$

   6) $Na \to Na^{1+} + e^-$ $F + e^- \to F^{1-}$

   7) $[Na^{1+}]\left[\text{⨯F⨯}\right]^{1-}$

   8)   NaF

   b) Mg   S
   1) $1s^22s^22p^63s^2$ $1s^22s^22p^63s^23p^4$

   2) ②   ⑥

   3) Mg: ⨯S⨯ (dot structure)

   4) lose 2   gain 2

   5) $[Mg^{2+}]$ $\left[\text{⨯S⨯}\right]^{2-}$

   6) $Mg \to Mg^{2+} + 2e^-$ $S + 2e^- \to S^{2-}$

   7) $[Mg^{2+}]\left[\text{⨯S⨯}\right]^{2-}$

   8)   MgS

c) Ca
1) $1s^2 2s^2 2p^6 3s^2 3p^6 4s^2$

2) ②

3) Ca:

4) lose 2

5) [Ca^{2+}]

6) Ca → Ca^{2+} + 2e$^-$

7) $\qquad\qquad$ [Ca^{2+}]$_3$ $\left[\begin{smallmatrix}x&x\\x&N&x\\\cdot&x\end{smallmatrix}\right]^{3-}_2$

8) $\qquad\qquad\qquad$ Ca$_3$N$_2$

N
$1s^2 2s^2 2p^3$

⑤

$\begin{smallmatrix}&xx&\\x&N&x\\&x&\end{smallmatrix}$

gain 3

$\left[\begin{smallmatrix}x&x&x\\x&N&x\\\cdot&x&\end{smallmatrix}\right]^{3-}$

N + 3e$^-$ → N^{3-}

d) Al
1) $1s^2 2s^2 2p^6 3s^2 3p^1$

2) ③

3) ·Al·

4) lose 3

5) [Al^{3+}]

6) Al → Al^{3+} + 3e$^-$

7) $\qquad\qquad$ [Al^{3+}] $\left[\begin{smallmatrix}x&xx&x\\\cdot&Br&x\\&xx&\end{smallmatrix}\right]^{1-}_3$

8) $\qquad\qquad\qquad$ AlBr$_3$

Br
$1s^2 2s^2 2p^6 3s^2 3p^6 4s^2 3d^{10} 4p^5$

⑦

$\begin{smallmatrix}&xx&\\x&Br&x\\&xx&\end{smallmatrix}$

gain 1

$\left[\begin{smallmatrix}x&xx&x\\\cdot&Br&x\\&xx&\end{smallmatrix}\right]^{1-}$

Br + e$^-$ → Br^{1-}

4. a) Li$_2$O
    b) Ba$_3$P$_2$
    c) Al$_2$S$_3$
    d) K$_3$N
    e) CaS
    f) AlP

Unit Three

5. a) I   I
    1) $1s^22s^22p^63s^23p^64s^23d^{10}4p^65s^24d^{10}5p^5$

    2) ⑦   ⑦

    3) ·Ï: ₓIₓ (xx top, xx bottom, x sides)

    4) gain 1   gain 1

    5) :Ï:ₓIₓ

    6) :Ï—Iₓ

    7) I_2

    b) O   O
    1) $1s^22s^22p^4$ $1s^22s^22p^4$

    2) ⑥   ⑥

    3) ·Ö: ₓOₓ

    4) gain 2   gain 2

    5) :Ö:ₓOₓ

    6) :Ö=Öₓ

    7) O_2

    c) N   N
    1) $1s^22s^22p^3$ $1s^22s^22p^3$

    2) ⑤   ⑤

    3) ·N̈·   ₓN̈ₓ

    4) gain 3   gain 3

INTRODUCTION TO CHEMISTRY

5) :N:ˣˣNˣˣ

6) :N≡Nˣˣˣˣ

7) N₂

d) Cl   Cl
1) $1s^2 2s^2 2p^6 3s^2 3p^5$ $1s^2 2s^2 2p^6 3s^2 3p^5$

2) ⑦   ⑦

3) ·C̈l: ˣˣC̈lˣˣˣ

4) gain 1   gain 1

5) :C̈l:ˣ C̈lˣˣˣˣ

6) :C̈l–C̈lˣˣˣˣ

7) Cl₂

6.   a) σ(1s – 4p)   b) σ(2pₓ – 2pₓ) and π(2p_y – 2p_y)

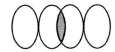

c) σ(2pₓ – 2pₓ) and π(2p_y – 2p_y) and π(2p_z – 2p_z)

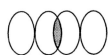

d) σ(1s – 2p)   e) σ(4p – 4p)

7. a) H :P: H   b) :Cl:
          ̈   :Cl:C:Cl:
         H   :Cl:

        H–P–H   :Cl:
          | :Cl–C–Cl:
          H   :Cl:

    c) :F:Be:F: d) H :Se: H

        :F—Be—F: H—Se—H

    e) :Cl:B:Cl: f) :O:S::O:
           :Cl: :O:

        :Cl–B–Cl: :O–S=O:
           :Cl: :O:

8. a) [:O:Br:O:]¹⁻   b) [:O:N::O:]¹⁻
           :O:

        [:O–Br–O:]¹⁻   [:O–N=O:]¹⁻
            |
           :O:

c) $\left[\begin{array}{c}:\ddot{O}:\overset{..}{S}:\ddot{O}:\\ :\ddot{O}:\end{array}\right]^{2-}$

$\left[\begin{array}{c}:\ddot{O}-S-\ddot{O}:\\ :\ddot{O}:\end{array}\right]^{2-}$

d) $\left[\begin{array}{c}:\ddot{O}:\\ :\ddot{O}:P:\ddot{O}:\\ :\ddot{O}:\end{array}\right]^{3-}$

$\left[\begin{array}{c}:\ddot{O}:\\ :\ddot{O}-P-\ddot{O}:\\ :\ddot{O}:\end{array}\right]^{3-}$

e) $\left[\begin{array}{c}H\\ H:\overset{..}{N}:H\\ H\end{array}\right]^{1+}$

$\left[\begin{array}{c}H\\ H-N-H\\ H\end{array}\right]^{1+}$

9. a) CCl$_4$
    1) AB$_4$
    2) tetrahedral
    3) tetrahedral
    4) 109.5°
    5) non-polar
    6)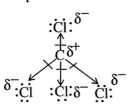

    b) PI$_3$
    1) AB$_3$E
    2) tetrahedral
    3) pyramidal
    4) less than 109.5°
    5) polar
    6)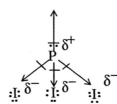

    c) H$_2$Se
    1) AB$_2$E$_2$
    2) tetrahedral
    3) bent or V-shaped
    4) less than 109.5°
    5) polar
    6)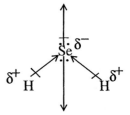

    d) BeF$_2$
    1) AB$_2$
    2) linear
    3) linear
    4) 180°
    5) non-polar
    6) $:\ddot{F}\overset{\delta^-}{\longleftarrow}\!\!+\!\!\overset{\delta^+}{Be}\!\!+\!\!\overset{\delta^-}{\longrightarrow}\ddot{F}:$

    e) BCl$_3$
    1) AB$_3$
    2) trigonal planar
    3) trigonal planar
    4) 120°
    5) non-polar
    6)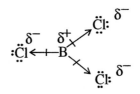

    f) SO$_3$
    1) AB$_3$
    2) trigonal planar
    3) trigonal planar
    4) 120°
    5) non-polar
    6)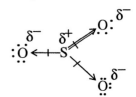

g) NO_2^{1-}
1) AB_2E
2) trigonal planar
3) bent or V-shaped
4) less than 120°
5) polar
6)

h) BrO_3^{1-}
1) AB_3E
2) tetrahedral
3) pyramidal
4) less than 109.5°
5) polar
6)

i) SO_3^{2-}
1) AB_3E
2) tetrahedral
3) pyramidal
4) less than 109.5°
5) polar
6)

j) PO_4^{3-}
1) AB_4
2) tetrahedral
3) tetrahedral
4) 109.5°
5) non-polar
6)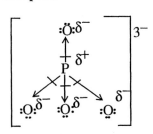

k) NH_4^+
1) AB_4
2) tetrahedral
3) tetrahedral
4) 109.5°
5) non-polar
6)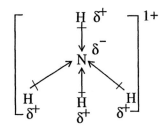

4 Unit Four

Chemical Calculations I

Unit Four

I. The atomic mass unit (amu)

II. The atomic weight of an element

A. Examples of atomic weight calculations

B. Counting the number of particles in an element

	C	S	Na
atomic wt.	12.01 amu		
gram atomic wt.	12.01 g		
# of atoms in a 1 GAW sample	6.022×10^{23}		

This number 6.022×10^{23} occurs so often it is given a special name: *Avogadro's number*

A sample containing 6.022×10^{23} fundamental building block units is said to contain 1 mole of units. The mole is analagous to the term "dozen".

The mass of 1 mole is called the molar mass

	C	S	Na
molar mass	12.01 g		

Examples of other molar masses

INTRODUCTION TO CHEMISTRY

Unit Four

C. **Converting moles of an element to grams: Solve by dimensional analysis or by formula application**

1. Example: What is the mass of 4.0 mol of C?

    By dimensional analysis: Restate problem as 4.0 mol C = ? g C

    Set-up unit factor $(4.0 \text{ mol C})\left(\dfrac{\text{g C}}{\text{mol C}}\right)$

    The mathematical relationship is determined by the molar mass: 1 mol = 12.01 g

    $(4.0 \text{ mol C})\left(\dfrac{12.01 \text{ g C}}{1 \text{ mol C}}\right) = 48.04 \text{ g C} = 48 \text{ g C}$

    By formula application: required formula is $\text{mol C} = \dfrac{\text{g C}}{\mathcal{M}\text{ C}}$

    Rearrange formula to solve for g C = (mol C)($\mathcal{M}$ C)

    g C = (4.0 mol)(12.01 g/mol) = 48.04g C = 48g C

2. More examples:

D. **Converting grams of an element to moles: by dimensional analysis or by formula application**

1. Example: 6.41g of sulfur contain how many moles of sulfur atoms

   By dimensional analysis: Re-state problem as 6.41g S = ? mol S

   Set up unit factor $(6.41 \text{ g S})\left(\dfrac{\text{mol S}}{\text{g S}}\right)$

   Determine mathematical relationship from the molar mass

   $(6.41 \text{ g S})\left(\dfrac{1 \text{ mol S}}{32.06 \text{ g S}}\right) = 0.200 \text{ mol S}$

   By formula application: required formula is $\text{mol S} = \dfrac{\text{g S}}{\mathcal{M} \text{ S}}$

   $\text{mol S} = \dfrac{6.41 \text{ g S}}{32.06 \text{ g S}/\text{mol S}} = 0.200 \text{ mol S}$

2. More examples:

E. **Converting moles of element to numbers of atoms**

1. Example: How many atoms in 4.0 mol of Al?
   By dimensional analysis: Re-state problem as 4.0 mol Al = ? atoms Al

   Set up unit factor $(4.0 \text{ mol Al}) \left(\dfrac{\text{atoms Al}}{\text{mol Al}} \right)$

   Determine mathematical relationship from mole definition: 1 mol = 6.022×10^{23} atoms

   $(4.0 \text{ mol Al}) \left(\dfrac{6.022 \times 10^{23} \text{ atoms Al}}{1 \text{ mol Al}} \right) = 24.088 \times 10^{23}$ atoms Al = 2.4×10^{24} atoms Al

   By formula application: The required formula is mol Al = $\dfrac{\text{atoms Al}}{\text{Avogadros \#}}$

   Rearrange this formula to give:
   atoms Al = (mol Al)(6.022×10^{23} atoms/mol)

   atoms Al = ((4.0 mol)(6.022×10^{23} atoms/mol) = 24.088×10^{23} atoms Al = 2.4×10^{24} atoms Al)

2. More examples

Unit Four

F. Converting number of atoms of an element to moles

1. Example: A sample of Mg contains 1.8×10^{22} atoms. How many mol of Mg are present?

   By dimensional analysis: Restate problem as 1.8×10^{22} atoms = ? mol Mg

   Set up unit factor $(1.8 \times 10^{22} \text{ atoms})\left(\dfrac{\text{mol Mg}}{\text{atoms Mg}}\right)$

   Determine mathematical relationship from mole definition: 1 mole = 6.022×10^{23} atoms

   $$(1.8 \times 10^{22} \text{ atoms Mg})\left(\dfrac{1 \text{ mol Mg}}{6.022 \times 10^{23} \text{ atoms Mg}}\right) = 0.030 \text{ mol Mg}$$

   By formula application: The required formula is $\text{mol Mg} = \dfrac{\text{atoms Mg}}{\text{Avogadros \#}}$

   $$\text{mol Mg} = \dfrac{1.8 \times 10^{22} \text{ atoms}}{6.022 \times 10^{23} \text{ atoms/mol}} = 0.030 \text{ mol Mg}$$

2. More examples:

Unit Four

G. Converting grams of an element to number of atoms

1. Example: 1.2g of C contains how many atoms of C?

    By dimensional analysis:

    $$(1.2 \text{g C})\left(\frac{1 \text{ mol C}}{12.0 \text{g C}}\right)\left(\frac{6.022 \times 10^{23} \text{ atoms C}}{1 \text{ mol C}}\right) = 6.0 \times 10^{22} \text{ atoms C}$$

    By formula application: $\text{mol C} = \dfrac{1.2 \text{g C}}{12.0 \text{ g/mol}} = 0.10 \text{ mol C}$,

    atoms C = $(0.10 \text{ mol})(6.022 \times 10^{23} \text{ atoms/mol}) = 6.0 \times 10^{22}$ atoms C

2. More examples:

H. Converting number of atoms to grams of an element

1. Example: 3.6×10^{23} atoms of Ca has what mass?

    By dimensional analysis: 3.6×10^{23} atoms Ca = ? g Ca

    $$(3.6 \times 10^{23} \text{ atoms Ca})\left(\frac{1 \text{ mol Ca}}{6.022 \times 10^{23} \text{ atoms Ca}}\right)\left(\frac{40.08 \text{g Ca}}{1 \text{ mol Ca}}\right) = 24 \text{g Ca}$$

    By formula application:

    $$\text{mol Ca} = \frac{3.6 \times 10^{23} \text{ atoms}}{6.022 \times 10^{23} \text{ atoms/mol}} = 0.60 \text{ mol Ca,}$$

    g Ca = $(0.60 \text{ mol})(40.88 \text{ g/mol}) = 24$ g Ca

2. More examples:

Unit Four

I. More calculations involving chemical compounds

1. Formula weight

	H_2O	CO_2	$C_6H_{12}O_6$	NaCl
Formula wt.				
Gram formula wt.				
# of sub-units in 1 GFW				
Molar mass				

J. Converting moles of a compound to grams

1. Example: 3.0 mol CO_2 has what mass?
   By dimensional analysis: 3.0 mol CO_2 = ? g CO_2

$$3.0 \text{ mol } CO_2 \left(\frac{44.0 \text{g } CO_2}{1 \text{ mol } CO_2}\right) = 132.0 \text{g } CO_2 = 1.3 \times 10^2 \text{g } CO_2$$

By formula application:

$$\text{g } CO_2 = (3.0 \text{ mol } CO_2)(4.40 \text{g/mol}) = 1.3 \times 10^2 \text{g } CO_2$$

2. More examples

Unit Four

K. **Converting grams of a compound to moles**

1. Example: 18.0g $C_6H_{12}O_6$ contains how many moles of molecules?
   By dimensional analysis:

   $$(18.0\text{g } C_6H_{12}O_6)\left(\frac{1 \text{ mol}}{180.0\text{g } C_6H_{12}O_6}\right) = 0.100 \text{ mol } C_6H_{12}O_6$$

   By formula application:

   $$\text{mol } C_6H_{12}O_6 = \frac{18.0\text{g } C_6H_{12}O_6}{180.0\text{g/mol}} = 0.100 \text{ mol } C_6H_{12}O_6$$

2. More examples:

L. **Convert moles of compound to # of building block units**

1. Example: 8.0 mol of NaCl contain how many ionic units
   By dimensional analysis: 8.0 mol NaCl = ? ionic units

   $$(8.0 \text{ mol NaCl})\left(\frac{6.022 \times 10^{23} \text{ ionic units}}{1 \text{ mol NaCl}}\right) = 4.8 \times 10^{24} \text{ ionic units}$$

   By formula application:

   ionic units = $(8.0 \text{ mol NaCl})(6.022 \times 10^{23} \text{ ionic units/mol}) = 4.8 \times 10^{24}$ ionic units

2. More examples:

INTRODUCTION TO CHEMISTRY

Unit Four

M. Convert # of building block units to moles of compound

1. Example: 1.2×10^{24} molecules H_2O equals how many moles H_2O?
   By dimensional analysis: 1.2×10^{24} molecules H_2O = ? mol H_2O

   $$(1.2 \times 10^{24} \text{ molecules } H_2O)\left(\frac{1 \text{ mol } H_2O}{6.022 \times 10^{23} \text{ molecules } H_2O}\right) = 2.0 \text{ mol } H_2O$$

   By formula application:

   $$\text{moles } H_2O = \frac{1.2 \times 10^{24} \text{ molecules } H_2O}{6.022 \times 10^{23} \text{ molecules/mol}} = 2.0 \text{ mol } H_2O$$

2. More examples:

N. Converting grams of compound to # of building block units

1. Example: 6.4 grams of CH_4 contains how many molecules?
   By dimensional analysis: 6.4g CH_4 = ? molecules CH_4

   $$(6.4 \text{g } CH_4)\left(\frac{1 \text{ mol } CH_4}{16.0 \text{g } CH_4}\right)\left(\frac{6.022 \times 10^{23} \text{ molecules } CH_4}{1 \text{ mol } CH_4}\right) = 2.4 \times 10^{23} \text{ molecules } CH_4$$

INTRODUCTION TO CHEMISTRY

By formula application: $\text{mol CH}_4 = \dfrac{6.4\text{g CH}_4}{16.0\text{g/mol}} = 0.40$ mol

$(0.40 \text{ mol CH}_4)(6.022 \times 10^{23} \text{ molecules/mol}) = 2.4 \times 10^{23}$ molecules CH_4

2. More examples:

O. **Converting # of building block units to grams of compound**
   1. Example: 3.0×10^{23} ionic units of $MgSO_4$ has what mass?

   By dimensional analysis: 3.0×10^{23} ionic units = ? g $MgSO_4$

   $(3.0 \times 10^{23} \text{ ionic units}) \left(\dfrac{1 \text{ mol MgSO}_4}{6.022 \times 10^{23} \text{ ionic units}} \right) \left(\dfrac{120.3 \text{ g MgSO}_4}{1 \text{ mol MgSO}_4} \right)$
   $= 6.0 \times 10^1$ g $MgSO_4$

   By formula application:

   $\text{mol MgSO}_4 = \dfrac{3.0 \times 10^{23} \text{ ionic units}}{6.022 \times 10^{23} \text{ ionic units/mol}} = 0.50$ mol $MgSO_4$

   g $MgSO_4$ = $(0.50 \text{ mol MgSO}_4)(120.3 \text{ g/mol}) = 6.0 \times 10^1$ g $MgSO_4$

   2. More examples:

Unit Four

P. Determining the percent composition of a compound

1. Example: Determine the % composition of H_2O
   Formula weight of H_2O = 18.0 amu

   $\%H = \dfrac{2.0}{18.0} \times 100 = 11\%$

   $\%O = \dfrac{16.0}{18.0} \times 100 = \underline{88.9\%}$

   $\phantom{\%O = \dfrac{16.0}{18.0} \times 100 = }$ 100% total

2. More examples:

Q. Empirical formula

1. Determining empirical formula from % composition by 4 step process
   (1) assume 100g of sample (2) convert to moles (3) Divide each mol value by the smallest
   (4) Convert to whole numbers (round-off criterion is ±0.05)
   Example: A sample is 85.6%C and 14.4%H. What is its empirical formula?

   $85.6\% \text{ C} \quad (85.6\text{g C})\left(\dfrac{1 \text{ mol C}}{12.01 \text{ g C}}\right) = 7.13 \text{ mol C}/7.13 = 1.00 \rightarrow 1$

   $14.4\% \text{ H} \quad (14.4\text{g H})\left(\dfrac{1 \text{ mol H}}{1.01 \text{ g H}}\right) = 14.3 \text{ mol H}/7.13 = 2.01 \rightarrow 2$

   CH_2

2. More examples:

INTRODUCTION TO CHEMISTRY

Unit Four

R. **Sometimes the values obtained in step 3 cannot be rounded off to a whole number.**

1. It must then be rounded off to a fraction involving small whole numbers. Suppose that the mole ratios of elements A and B after step 3 are

$$\left.\begin{array}{ll} A & 1.00 \to 1 \times 2 = 2 \\ B & 1.48 \to 1.5 \times 2 = 3 \end{array}\right\} A_2B_3$$

2. More examples:

S. **Knowing the empirical formula and the molar mass allows the determination of the chemical formula**

1. Example: empirical formula = CH_2, molar mass = 42 g/mol
The true chemical formula will be a multiple of the empirical formula $(CH_2)_n$
n can be determined from the molar mass and the empirical weight
empirical weight

INTRODUCTION TO CHEMISTRY

$$n = \frac{\text{molar mass}}{\text{empirical wt.}} \rightarrow \text{rounded-off to the nearest whole number}$$

$$n = \frac{42}{14} = 3.0 \rightarrow 3$$

chemical formula = $(CH_2)_3$ = C_3H_6

2. More examples:

Summary of Calculations

Flow chart for dimensional analysis:

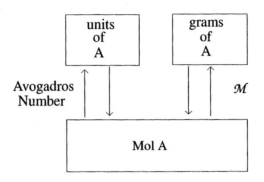

Rules for using flow chart
1. Going up the chart -

2. Going down the chart -

Formulas:

$$\text{mol A} = \frac{\text{grams A}}{\text{molar mass A}}$$
$$\downarrow$$
$$\text{molar mass A} =$$
$$\downarrow$$
$$\text{grams A} =$$

$$\text{moles A} = \frac{\text{building block units of A}}{6.022 \times 10^{23}}$$
$$\downarrow$$
$$\text{building block units of A} =$$

Steps to calculate empirical formula from % composition
(1) Assume 100g of sample
(2) Convert grams to moles
(3) Divide each mole by the smallest
(4) Round-off to whole numbers. Round-off to fractions if necessary first.
If molar mass is known determine chemical formula from

$$n = \frac{\text{molar mass}}{\text{empirical wt.}} \qquad \text{chemical formula} = (\text{empirical formula})_n$$

Unit Four

Things to Know

Definitions
1. atomic mass unit
2. atomic weight
3. gram atomic weight
4. Avogadro's number
5. mole
6. molar mass
7. formula weight
8. gram formula weight
9. percent composition
10. empirical formula
11. empirical weight

Concepts

1. Be able to calculate atomic weight from atomic masses and percent abundances.
2. Be able to determine atomic weight, gram atomic weight and molar mass from periodic table for a given element.
3. Be able to convert moles to grams for elements and compounds.
4. Be able to convert grams to moles for elements and compounds.
5. Be able to convert moles to building block units for elements and compounds
6. Know the building block units of elements, covalent compounds, and ionic compounds.
7. Be able to convert building block units to moles for elements and compounds.
8. Be able to determine formula weight, gram formula weight, and molar mass of a compound.
9. Be able to convert grams to building block units for elements and compounds.
10. Be able to convert building block units to grams for elements and compounds.
11. Be able to determine % composition of a compound.
12. From percent composition be able to determine empirical formula.
13. From empirical formula and molar mass be able to determine the chemical formula.

Review Questions

1. The mass of the _____ atom is the basis for the atomic mass unit.
2. The symbol for the atomic mass unit is the _____.
3. The weighted average of the atomic masses of all naturally occuring isotopes of an element is the _____.
4. The atomic weight measured in grams is the _____ and is abbreviated by _____.
5. The value of Avogadros number is _____.
6. A sample of an element containing Avogadro's number of atoms is called a _____.
7. The mass of one mole of an element is called its _____.
8. The sum of the atomic weights of all atoms or ions in a compound is its _____.
9. The formula weight measured in grams is the _____.
10. The building block unit for a covalent compound is the _____.
11. A sample of a covalent compound consisting of Avogadro's number of molecules is called a _____.
12. The mass of one mole of a covalent compound is called its _____.
13. Another name for the molar mass of a covalent compound is _____.
14. The building block unit for an ionic compound is the _____.
15. A sample of an ionic compound containing Avogadro's number of ionic units is called a _____.
16. The mass of one mole of an ionic compound is called its _____.
17. Another name for the molar mass of an ionic compound is _____.
18. The sum of the percentages in the percent composition of a compound must be _____.
19. An empirical formula gives the types of atoms and the lowest _____ ratio of atoms present in a compound.
20. In determining empirical formula from percent composition step 1 is to assume _____ grams of sample, step 2 is to convert grams to _____, step 3 is to divide each mole value by the _____, step 4 is to round-off to _____.
21. In the process described in #20 the round-off criterion is _____.
22. If it is not possible to round-off directly to a whole number it must be rounded-off to a _____ involving _____ whole numbers.
23. To get from empirical formula to chemical formula the _____ must be known.

Review Problems

1. Determine the atomic weight of each element to 4 significant digits.
   a) Ca
   b) Ni
   c) Br
   d) Ba
   e) Ag
   f) Pt
   g) U

2. Determine the molar mass of each element to 4 significant digits.
   a) Ca
   b) Ni
   c) Br
   d) Ba
   e) Ag
   f) Pt
   g) U

3. Convert each mole value to grams.
   a) 2.70 mol Na
   b) 4.00 mol Sr
   c) 0.20 mol S
   d) 0.50 mol Ar
   e) 4.00×10^{-3} mol Pb
   f) 2.000×10^{-2} mol Hg

4. Convert each gram value to moles.
   a) 6.54g Zn
   b) 41.8g Bi
   c) 3.024g H
   d) 1.002g Ca
   e) 9.6g S
   f) 0.24g U

5. Convert each mole value to number of atoms.
   a) 5.0 mol Al
   b) 246 mol Fe
   c) 0.40 mol Cu
   d) 0.006 mol Na
   e) 2.000 mol S
   f) 8.000×10^{-4} mol C

6. Convert each quantity of atoms to moles.
   a) 3.0×10^{26} atoms K
   b) 6.0×10^{21} atoms Ba
   c) 9.03×10^{24} atoms B
   d) 1.00×10^{20} atoms Be
   e) 1.204×10^{27} atoms P
   f) 1.800×10^{10} atoms N

7. Convert each gram value to number of atoms.
   a) 0.31g P
   b) 6.42×10^{-3} g S
   c) 6.054×10^{2} g Ne

Unit Four

8. Convert each quantity of atoms to grams.
   a) 3.01×10^{22} atoms I
   b) 8.0×10^{25} atoms Se
   c) 3.011×10^{24} atoms Kr

9. Determine the formula weight of each compound to 4 significant digits.
   a) Cl_2O_7
   b) $C_{12}H_{22}O_{11}$
   c) $HC_2H_3O_2$
   d) $Al_2(SO_4)_3$
   e) $Mg(NO_3)_2$
   f) $Ca_3(PO_4)_2$

10. Determine the molar mass of each compound to 4 significant digits.
    a) Cl_2O_7
    b) $C_{12}H_{22}O_{11}$
    c) $HC_2H_3O_2$
    d) $Al_2(SO_4)_3$
    e) $Mg(NO_3)_2$
    f) $Ca_3(PO_4)_2$

11. For each compound decide if the building block unit is the molecule or the ionic unit.
    a) Cl_2O_7
    b) $C_{12}H_{22}O_{11}$
    c) $HC_2H_3O_2$
    d) $Al_2(SO_4)_3$
    e) $Mg(NO_3)_2$
    f) $Ca_3(PO_4)_2$

12. Convert each mole value to grams.
    a) 8.00 mol CO_2
    b) 3.00×10^{-2} mol KBr
    c) 4.0×10^{3} mol CCl_4

13. Convert each gram value to moles.
    a) 3.2×10^{-2} g SO_2
    b) 68.0g $CaSO_4$
    c) 6.4g O_2

14. Convert moles to molecules or ionic units.
    a) 2.0 mol CH_4
    b) 4.00×10^{-3} mol HNO_3
    c) 8.35 mol NH_3

15. Convert molecules or ionic units to moles.
    a) 1.2×10^{24} ionic units Al_2S_3
    b) 5.600×10^{23} molecules N_2
    c) 3.01×10^{22} ionic units K_2O

16. Convert each gram value to molecules or ionic units.
    a) 1.8g $C_6H_{12}O_6$
    b) 1.7×10^{-2} g $NaNO_3$
    c) 684g $C_{12}H_{22}O_{11}$

17. Convert molecules or ionic units to grams.
    a) 1.00×10^{24} molecules HCl
    b) 2.4×10^{25} ionic units K_3N
    c) 7.5×10^{21} molecules Br_2

18. Determine the percent composition of each.
    a) CO_2
    b) SO_3
    c) CH_4
    d) $CaSO_4$
    e) Al_2S_3
    f) $Mg(NO_3)_2$

19. Determine the empirical formula of each.
    a) $C_6H_{12}O_6$
    b) N_2H_4
    c) $C_{12}H_{22}O_{11}$
    d) Cl_2O_7
    e) P_4O_{10}
    f) C_6H_6

20. Determine the empirical formula from each percent composition.
    a) %S = 50.05%
       %O = 49.95%
    b) %H = 2.06%
       %S = 32,69%
       %O = 65.25%
    c) %Fe = 69.94%
       %O = 30.06%
    d) %C = 82.63%
       %H = 17.4%

Answers to Review Problems

1. a) 40.08 amu b) 58.71 amu c) 79.90 amu
   d) 137.3 amu e) 107.9 amu f) 195.1 amu
   g) 238.0 amu

2. a) 40.08g b) 58.71g c) 79.90g
   d) 137.3g e) 107.9g f) 195.1g
   g) 238.0g

3. a) 62.1g Na b) 3.50×10^2 g Sr c) 6.4g S
   d) 2.0×10^1 g Ar e) 0.829g Pb f) 4.012g Hg

4. a) 0.100 mol Zn b) 0.0200 mol Bi c) 3.000 mol H
   d) 0.02500 mol Ca e) 0.30 mol S f) 1.0×10^{-3} mol U

5. a) 3.0×10^{24} atoms Al
   b) 1.48×10^{26} atoms Fe
   c) 2.4×10^{23} atoms Cu
   d) 4×10^{21} atoms Na
   e) 1.204×10^{24} atoms S
   f) 4.818×10^{20} atoms C

6. a) 5.0×10^2 mol K b) 0.010 mol Ba c) 15.0 mol B
   d) 1.66×10^{-4} mol Be
   e) 1.999×10^3 mol P
   f) 2.989×10^{-14} mol N

7. a) 6.0×10^{21} atoms P
   b) 1.21×10^{20} atoms S
   c) 1.807×10^{25} atoms Ne

8. a) 6.34g I b) 1.0×10^4 g Se c) 419.0g Kr

9. a) 182.9 amu b) 342.3g c) 60.05 amu
   d) 342.1 amu e) 148.3 amu f) 310.2 amu

Unit Four

10. a) 182.9g　b) 342.3g　c) 60.05g
    d) 342.1g　e) 148.3g　f) 310.2g

11. a) molecule　b) molecule　c) molecule
    d) ionic unit　e) ionic unit　f) ionic unit

12. a) 352g CO_2　b) 3.57g KBr　c) 6.2×10^5 g CCl_4

13. a) 5.0×10^{-4} mol SO_2
    b) 0.500 mol $CaSO_4$
    c) 0.20 mol O_2

14. a) 1.2×10^{24} molecules CH_4
    b) 2.41×10^{21} molecules HNO_3
    c) 5.03×10^{24} molecules NH_3

15. a) 2.0 mol Al_2S_3　b) 0.9299 mol N_2　c) 0.0500 mol K_2O

16. a) 6.0×10^{21} molecules $C_6H_{12}O_6$
    b) 1.2×10^{20} ionic units $NaNO_3$
    c) 1.20×10^{24} molecules $C_{12}H_{22}O_{11}$

17. a) 60.5g HCl　b) 5.2×10^3 g K_3N　c) 2.0g Br_2

18. a) %C = 27.29%　b) %S = 40.04%　c) %C = 74.83%
    %O = 72.71%　　%O = 59.96%　　%H = 25.2%
    d) %Ca = 29.44%
    %S = 23.55%
    %O = 47.01%
    e) %Al = 35.94%
    %S = 64.06%
    f) %Mg = 16.38%
    %N = 18.89%
    %O = 64.72%

19. a) CH_2O　b) NH_2　c) $C_{12}H_{22}O_{11}$
    d) Cl_2O_7　e) P_2O_5　f) CH

20. a) SO_2　b) H_2SO_4　c) Fe_2O_3
    d) C_2H_5

INTRODUCTION TO CHEMISTRY

5 Unit Five

Chemical Reactions and Stoichiometry

I. **Reactant**

II. **Products**

III. **Symbolism in chemical reactions**

    A. Reactants → Products

        1. (s)

        2. (*l*)

        3. (g)

        4. (aq)

        5. $\xrightarrow{\Delta}$

    B. **In order to satisfy the Law of Conservation of Mass chemical reactions must be balanced**

        1. Example:

$$\_\_N_2(g) + \_\_H_2(g) \rightarrow \_\_NH_3(g)$$
            2 N atoms   2 H atoms   1 N atom
                                               3 H atoms

$$\_\_N_2(g) + \_\_H_2(g) \rightarrow \underline{2}\ NH_3(g)$$
            2 N atoms   2 H atoms   2 N atom
                                               6 H atoms

$$\_\_N_2(g) + \underline{3}\ H_2(g) \rightarrow \underline{2}\ NH_3(g)$$
            2 N atoms   6 H atoms   2 N atom
                                               6 H atoms

        2. More examples:

Unit Five

IV. Types of Chemical Reactions

A. Combination or Synthesis: $A + B \rightarrow C$
1. Examples:

B. Decomposition: $AB \rightarrow A + B$
1. Examples:

C. Single replacement: $A + BC \rightarrow B + AC$
1. Examples:

D. Double replacement or metathesis: $AC + BD \rightarrow AD + BC$
1. Examples:

E. Neutralization: Acid + Base $\rightarrow$ Salt + H_2O
1. Examples:

F. **In single replacement reactions the more active metal will replace the less active metal.**
1. Activity series:

2. Examples:

G. **For a double replacement reaction to occur:**
1. both reactants must be (aq)
2. at least one product must be (s), (*l*), or (g)

H. **Solubility Rules**

I. **Solutions**
1. Solutions have two components: solute and solvent

solute vs. solvent

Unit Five

Unit Five

J. Electrical properties of solutions
1. Good conductors
2. Poor conductors
3. Non-conductors

K. Solutes can be classified as:
1. Strong electrolytes
2. Weak electrolytes
3. Non-electrolytes

L. Electrolytes undergo either dissociation or ionization in aqueous solution

Dissociation vs. Ionization

1. Examples of dissociation

2. Examples of ionization

M. Molecular Equation for metathesis reactions:

Unit Five

N. Converting molecular equation to complete ionic equation:

O. Spectator ions

P. Converting complete ionic equation to net ionic equation:

More examples of converting molecular equation to net ionic form

Q. Interpreting the balanced coefficients of a chemical reaction

Unit Five

R. **Mole—mole stoichiometry problems**

1. Example:
Consider the reaction $N_2(g) + 3H_2(g) \rightarrow 2NH_3(g)$. If 2.0 mol $N_2(g)$ reacts with excess $H_2(g)$ how many moles of $NH_3(g)$ can be produced?
By dimensional analysis: 2.0 mol $N_2(g)$ = ? mol $NH_3(g)$

$$2.0 \text{ mol } N_2 \left(\frac{2 \text{ mol } NH_3}{1 \text{ mol } N_2} \right) = 4.0 \text{ mol } NH_3$$

By formula application:

$$\frac{1}{a}(\text{mol A}) = \frac{1}{b}(\text{mol B})$$

$$\frac{1}{1} \text{ mol } N_2 = \frac{1}{2} \text{ mol } NH_3 \rightarrow 2.0 = \frac{1}{2} \text{ mol } NH_3 \rightarrow \text{mol } NH_3 = 2.0(2) = 4.0 \text{ mol}$$

2. More examples:

S. **Mass-mole and mole-mass problems**

1. Consider the reaction: $N_2(g) + 3H_2(g) \rightarrow 2NH_3(g)$. How many grams of $NH_3(g)$ can be produced when 9.00 mol H_2 reacts with excess N_2?
By dimensional analysis:

$$9.00 \text{ mol } H_2 \left(\frac{2 \text{ mol } NH_3}{3 \text{ mol } H_2} \right) \left(\frac{17.0 \text{g } NH_3}{1 \text{ mol } NH_3} \right) = 102 \text{g } NH_3$$

INTRODUCTION TO CHEMISTRY

By formula application:

$$\frac{1}{3}(9.00 \text{ mol } H_2) = \frac{1}{2} \text{ mol } NH_3 \rightarrow \text{mol } NH_3 = 6.00 \text{ mol}$$

$$\text{gram } NH_3 = (6.00 \text{ mol})(17.0 \text{g/mol}) = 102 \text{g } NH_3$$

2. More examples:

T. **Mass—mass problems**
   1. Example:
   Consider the reaction $N_2(g) + 3H_2(g) \rightarrow 2NH_3(g)$. How many grams of $NH_3(g)$ can be produced when 5.6g N_2 reacts with excess H_2.
   By dimensional analysis:

   $$5.6 \text{g } N_2 \left(\frac{1 \text{ mol } N_2}{28.0 \text{g } N_2}\right)\left(\frac{2 \text{ mol } NH_3}{1 \text{ mol } N_2}\right)\left(\frac{17.0 \text{g } NH_3}{1 \text{ mol } NH_3}\right) = 6.8 \text{g } NH_3$$

By formula application:

$$\frac{1}{1}\left(\frac{5.6g}{28.0 g/mol}\right) = \frac{1}{2}\left(\frac{g\ NH_3}{17.9 g/mol}\right) \rightarrow g\ NH_3 = 6.8g\ NH_3$$

2. More examples

U. **theoretical yield**

V. **actual yield**

W. **%yield** = $\frac{\text{actual yield}}{\text{theoretical yield}} \times 100$

1. Example: A certain reaction has theoretical yield = 20.0g with actual yield = 8.00g
   What is the percent yield?

   %yield = $\frac{8.00g}{20.0g} \times 100 = 40.0\%$

   What happens when specific amounts of each reactant are given?
   Limiting reagent (reactant)

2. Non-chemical example

Unit Five

X. **Limiting reagent calculation for the general reaction** $aA + bB \rightarrow$ Products

1. Ratio comparison method: Compare $\dfrac{\text{mol A}}{\text{mol B}}$ to $\dfrac{a}{b}$

   If $\dfrac{\text{mol A}}{\text{mol B}} < \dfrac{a}{b}$ then A is the limiting reagent.

   If $\dfrac{\text{mol A}}{\text{mol B}} > \dfrac{a}{b}$ then B is the limiting reagent.

   If $\dfrac{\text{mol A}}{\text{mol B}} = \dfrac{a}{b}$ then A and B are both limiting reagents.

2. Theoretical yield method:
   Calculate product yield from mol A
   Calculate product yield from mol B
   Whichever quantity gives the lowest product theoretical yield is the limiting reactant.

3. Example: Consider the reaction $N_2(g) + 3H_2(g) \rightarrow 2NH_3(g)$. If 3.00 mol N_2 reacts with 10.0 mol H_2 which reactant is the limiting reagent? What is the theoretical yield of NH_3?

   a) By ratio comparison:

   $\dfrac{3.00 \text{ mol } N_2}{10.0 \text{ mol } H_2} < \dfrac{1}{3}$ so the limiting reagent is N_2.

   Using the limiting reagent:

   $3.00 \text{ mol } N_2 \left(\dfrac{2 \text{ mol } NH_3}{1 \text{ mol } N_2}\right)\left(\dfrac{17.0 \text{g } NH_3}{1 \text{ mol } NH_3}\right) = 102 \text{g } NH_3$

   b) By theoretical yield method:

   $3.00 \text{ mol } N_2 \left(\dfrac{2 \text{ mol } NH_3}{1 \text{ mol } N_2}\right)\left(\dfrac{17.0 \text{g } NH_3}{1 \text{ mol } NH_3}\right) = 102 \text{g } NH_3$

   $10.0 \text{ mol } H_2 \left(\dfrac{2 \text{ mol } NH_3}{3 \text{ mol } H_2}\right)\left(\dfrac{17.0 \text{g } NH_3}{1 \text{ mol } NH_3}\right) = 113 \text{g } NH_3$

   $\rightarrow N_2$ is the limiting reagent since it gives the smallest theoretical yield

4. More examples

Flowchart Summary for Stoichiometry Calculations

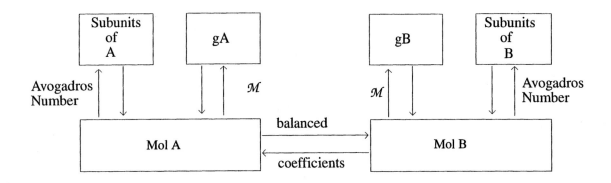

Things to Know

1. Reactant
2. Product
3. aqueous
4. combination reaction
5. decomposition reaction
6. single replacement reaction
7. double replacement reaction
8. neutralization reaction
9. solute
10. solvent
11. strong electrolyte
12. weak electrolyte
13. non-electrolyte
14. dissociation
15. ionization
16. molecular equation
17. complete ionic equation
18. spectator ion
19. net ionic equation
20. actual yield
21. theoretical yield
22. percent yield
23. limiting reagent

Concepts

1. Know the symbolism used in a chemical reaction.
2. Be able to balance chemical reactions.
3. Know the five types of chemical reactions.
4. Given a reaction be able to classify it.
5. Be able to use the activity series to decide if a single replacement reaction will occur.
6. Know the criteria for a metathesis reaction to occur.
7. Know the solubility rules.
8. Know the electrical properties of solutions.
9. For a metathesis reaction be able to convert molecular equation to net ionic equation and identify spectator ions.
10. Be able to work mole–mole stoichiometry problems.
11. Be able to work mole–mass and mass–mole problems.
12. Be able to work mass–mass problems.
13. Be able to calculate percent yield.
14. Be able to determine the limiting reagent.

Review Questions

1. The substances combined together in a chemical reaction are called _____.
2. The substances produced in a chemical reaction are called _____.
3. The → in a chemical reaction is read "_____" or "_____."
4. A reactant or product in the solid state is represented by placing _____ after the formula, in the liquid state a _____ is used, in the gaseous state a _____ is used, dissolved in H_2O a _____ is used.
5. A "Δ" above the arrow indicates that _____ is required.
6. According to the Law of Conservation of mass chemical reactions must be _____.
7. A balanced reaction means that the number of _____ of each type of element must be the same as reactant and product.
8. What is the symbolic representation of a synthesis reaction? _____
9. What is the symbolic representation of a decomposition reaction? _____
10. What is the symbolic representation of a single replacement reaction? _____
11. What is the symbolic representation of a metathesis reaction? _____
12. What is the symbolic representation of a neutralization reaction? _____
13. In single replacement reactions the _____ active metal will replace the active metal.
14. For a metathesis reaction to occur both reactants must be _____ and at least one product must be _____, _____, or _____.
15. All Na^+, K^+ and NH_4^+ compounds are _____.
16. All nitrates, chlorates, perchlorates and acetates are _____.
17. Most chlorides, bromides, and iodides are _____ except when combined with _____, _____, _____
18. Most sulfates are _____ except when combined with _____, _____, _____.
19. Most phosphates, carbonates, hydroxides, sulfides and fluorides are _____ except when combined with _____, _____, _____.
20. A homogeneous mixture is also called a _____.
21. The substance dissolved in a solution is the _____.
22. The dissolving medium in a solution is the _____.
23. To be a good conductor there must be a _____ concentration of ions in solution.
24. A _____ concentration of ions in solution makes it a poor conductor.
25. A solution which is a non-conductor has a _____ concentration of ions in solution.
26. A solute which produces a 100% yield of ions in solution is a _____ electrolyte.
27. A solute which produces much less than a 100% yield of ions in solution is a _____ electrolyte.
28. A solute which produces a 0% yield of ions in solution is a _____ electrolyte.

29. Pure water is a _____ electrolyte.
30. When an ionic compound dissolves it undergoes _____.
31. Dissociation is the _____ of ions which are already present.
32. When a covalent compound dissolves it may undergo _____.
33. Ionization is the _____ of ions where there were none.
34. Ions which do not participate in a reaction but are present in solution are called _____ ions.
35. The calculated value for the mass of product produced is called the _____ yield.
36. The amount of product produced in the laboratory is called the _____ yield.
37. The reactant which is completely consumed in a reaction is called the _____.

Review Problems

1. Balance each of the following reactions.
   a) $Fe(s) + O_2(g) \rightarrow Fe_2O_3(s)$
   b) $KClO_3(s) \rightarrow KCl(s) + O_2(g)$
   c) $C_2H_6(g) + O_2(g) \rightarrow CO_2(g) + H_2O(g)$
   d) $AgNO_3(aq) + MgCl_2(aq) \rightarrow Mg(NO_3)_2(aq) + AgCl(s)$
   e) $P_4O_{10}(g) + H_2O(l) \rightarrow H_3PO_4(aq)$
   f) $C_2H_6O(l) + O_2(g) \rightarrow CO_2(g) + H_2O(g)$

2. Classify each reaction by type.
   a) $2SO_2(g) + O_2(g) \rightarrow 2SO_3(g)$
   b) $H_2CO_3(aq) \rightarrow CO_2(g) + H_2O(g)$
   c) $Mg(s) + 2HCl(aq) \rightarrow MgCl_2(aq) + H_2(g)$
   d) $NaOH(aq) + HNO_3(aq) \rightarrow NaNO_3(aq) + H_2O(l)$
   e) $Ba(NO_3)_2(aq) + K_2SO_4(aq) \rightarrow 2KNO_3(aq) + BaSO_4(s)$
   f) $2Na(s) + 2H_2O(l) \rightarrow 2NaOH(aq) + H_2(g)$

3. Decide which of the following single replacement reactions will occur.
   a) $Cu(s) + HCl(aq) \rightarrow$
   b) $Mg(s) + NaOH(aq) \rightarrow$
   c) $Al(s) + CuSO_4(aq) \rightarrow$
   d) $Fe(s) + HCl(aq) \rightarrow$
   e) $Na(s) + Mg(NO_3)_2(aq) \rightarrow$
   f) $K(s) + H_2O(l) \rightarrow$

4. Decide on the solubility of each compound.
   a) Na_3PO_4
   b) $Al(NO_3)_3$
   c) $Ba(OH)_2$
   d) $CaCO_3$
   e) NH_4OH
   f) $PbSO_4$
   g) $AgCl$
   h) K_2S
   i) $HgSO_4$

5. Convert each molecular equation into complete ionic and net ionic reactions. Identify spectator ions.
   a) $3AgNO_3(aq) + AlCl_3(aq) \rightarrow Al(NO_3)_3(aq) + 3AgCl(s)$
   b) $Na_2SO_4(aq) + Ba(ClO_3)_2(aq) \rightarrow 2NaClO_3(aq) + BaSO_4(s)$

6. Consider the reaction:
   $2C_4H_{10}(l) + 13O_2(g) \rightarrow 8CO_2(g) + 10H_2O(g)$.
   a) If 8.00 mol C_4H_{10} reacts with excess O_2 how many moles of H_2O can be produced?
   b) If 2.60 mol O_2 reacts with excess C_4H_{10} how many grams of CO_2 can be produced?
   c) If 11.6g C_4H_{10} reacts with excess O_2 how many grams of H_2O can be produced?
   d) If 5.8g C_4H_{10} reacts with 16.0g O_2 identify the limiting reagent.

e) Continuing from (d) calculate the theoretical yield of $CO_2(g)$.
f) Continuing from (d) calculate the grams of excess reactant.
g) If the actual yield of CO_2 is 7.65g calculate the % yield.

7. Consider the reaction: $4NH_3(g) + 5O_2(g) \rightarrow 4NO(g) + 6H_2O(g)$
a) If 1.20 mol NH_3 reacts with excess O_2 how many moles of H_2O can be produced?
b) If 2.50 mol O_2 reacts with excess NH_3 how many grams of NO can be produced?
c) If 3.40g of NH_3 reacts with excess O_2 how many grams of H_2O can be produced?
d) If 8.50g NH_3 reacts with 25.6g O_2 identify the limiting reagent.
e) Continuing from (d) calculate the theoretical yield of NO.
f) Continuing from (d) calculate the grams of excess reactant.
g) If the actual yield is 10.0g calculate the % yield.

Answers to Review Problems

1. a) $4 + 3 \rightarrow 2$ c) $2 + 7 \rightarrow 4 + 6$ e) $1 + 6 \rightarrow 4$
    b) $2 \rightarrow 2 + 3$ d) $2 + 1 \rightarrow 1 + 2$ f) $1 + 3 \rightarrow 2 + 3$

2. a) synthesis
    b) decomposition
    c) single replacement
    d) neutralization
    e) metathesis
    f) single replacement

3. a) no reaction   c) reaction   e) reaction
    b) no reaction   d) reaction   f) reaction

4. a) soluble   d) not soluble   g) not soluble
    b) soluble   e) soluble   h) soluble
    c) not soluble   f) not soluble   i) not soluble

5. a) $3Ag^+(aq) + 3NO_3^{1-}(aq) + Al^{3+}(aq) + 3Cl^{1-}(aq) \rightarrow$
    $Al^{3+}(aq) + 3NO_3^{1-}(aq) + 3AgCl(s)$
    $Ag^+(aq) + Cl^-(aq) \rightarrow AgCl(s)$
    spectator ions: Al^{3+}, NO_3^{1-}

    b) $2Na^{1+}(aq) + SO_4^{2-}(aq) + Ba^{2+}(aq) + 2ClO_3^{1-}(aq) \rightarrow$
    $2Na^{1+}(aq) + 2ClO_3^{1-}(aq) + BaSO_4(s)$
    $Ba^{2+}(aq) + SO_4^{2-}(aq) \rightarrow BaSO_4(s)$
    spectator ions: Na^{1+}, ClO_3^{1-}

6. a) 40.0 mol H_2O d) O_2 g) 56.7%
    b) 70.4g CO_2 e) 13.5g CO_2
    c) 18.0g H_2O f) 1.3g C_4H_{10}

7. a) 1.80 mol H_2O d) NH_3 g) 66.7%
    b) 60.0g NO   e) 15.0g NO
    c) 5.40g H_2O f) 5.6g

6 Unit Six

Gases, Liquids, and Solids

I. Properties of Gases

II. State Variables

   A. Temperature

   B. Volume

   C. Moles

   D. Pressure

      1. SI units

      2. common units

III. Gas Laws

   A. Boyle's Law

      1. Word description

      2. Formula

      3. Examples

Unit Six

B. **Charles' Law**
   1. Word description

   2. Formula

   3. Examples

C. **The universal gas law or ideal gas law**
   1. Formula
   2. Examples

Ideal gas vs Real gas

D. **Dalton's Law of Partial Pressures**
   1. Word description

   2. Formula

   3. Examples

E. The kinetic molecular theory of gases

F. Using KMT to explain the gas laws

IV. Properties of Liquids

A. The kinetic molecular theory of liquids

Unit Six

B. **The vaporization of liquids**
   1. In an open container

   2. In a closed container

C. **Equilibria can be either static or dynamic**

   static equilibrium vs. dynamic equilibrium

D. **Equilibrium vapor pressure**

   volatile liquids vs non-volatile liquids

E. **The boiling of liquids and the stability of the vapor bubbles**
   1. When the external pressure = 1atm
      Pascal's Law
      a) At a T for which v.p. < 1atm the vapor bubbles will collapse

INTRODUCTION TO CHEMISTRY

b) At a T for which v.p. = 1atm the vapor bubbles will be stable

Normal boiling point
boiling point

2. When the external pressure ≠ 1atm the liquid will boil at a temperature either lower or higher than the normal boiling point

   a) At a T for which v.p. < external pressure

   b) At a T for which v.p. = external pressure

F. **Properties of Solids**

Unit Six

G. **Phase diagrams—summary of the state of a substance at various P and T**

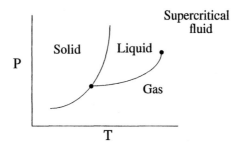

1. Triple point
2. Critical point consists of the critical T and the critical P

3. Supercritical fluid

H. **Unusual properties of water**
   1. The density of liquid water reaches its maximum at a temperature above the freezing point of water.

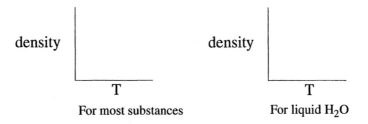

   2. The density of solid water is less than that of liquid water.

   3. Under extremely high pressure ice melts.

Phase diagram for H₂O

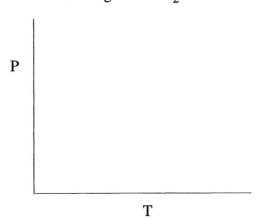

I. **The properties of liquids are determined by its intermolecular forces**

J. **Types of intermolecular forces**

   1. Dipole—dipole

   2. Hydrogen bonding

3. London dispersion

K. There are other inter-particle forces involving ions

L. Forces acting between molecules of a polar liquid

M. Forces acting between molecules of a non-polar liquid

N. The stronger the intermolecular forces the
1. higher the melting point

2. higher the boiling point

3. lower the vapor pressure at a given temperature

4. higher the surface tension
    surface tension

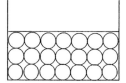

Adhesion vs Cohesion

O. The wetting of surfaces
   1. If adhesion > cohesion

   2. If adhesion < cohesion

Things to Know

Definitions
1. Pascal
2. Atmosphere
3. Torr
4. mm Hg
5. Boyle's Law
6. Charles' Law
7. Universal Gas Law
8. Dalton's Law of Partial Pressures
9. static equilibrium
10. dynamic equilibrium
11. vapor pressure
12. volatile
13. normal boiling point
14. boiling point
15. triple point
16. critical temperature
17. critical pressure
18. supercritical fluid
19. dipole—dipole forces
20. hydrogen bonding forces
21. london dispersion forces
22. surface tension
23. adhesion
24. cohesion

Concepts

1. Know the properties of gases.
2. Be able to convert between the pressure units.
3. Be able to solve Boyle's Law problems.
4. Be able to solve Charles' Law problems.
5. Be able to solve Universal Gas Law problems.
6. Be able to solve Dalton's Law of Partial Pressure problems.
7. Know the postulates of the kinetic molecular theory.
8. Know the properties of liquids.
9. Know the difference in the vaporization of a liquid in an open container vs a closed container.
10. Know the conditions for a liquid to be at its normal boiling point.
11. Know the conditions for a liquid to be at its boiling point.

12. Know the properties of solids.
13. Know the various parts of a typical phase diagram.
14. Know the unusual properties of water.
15. Know the types of intermolecular forces.
16. Know the properties of each type of intermolecular force.
18. Given a polar substance be able to predict the intermolecular forces acting.
19. Given a non-polar substance be able to predict the intermolecular forces acting.
20. Know how the strength of the intermolecular forces determine the melting point, boiling point, vapor pressure, and surface tension.
21. Know how the relative strengths of adhesion and cohesion determines the wetting of surfaces.

Review Questions

1. Gases have _____ shape or volume.
2. Gases take the _____ of their container and completely _____.
3. Gases have _____ density measured in _____.
4. Gases _____ and exhibit _____.
5. The state of a gas is completely determined by its _____.
6. _____ is a man-made scale to measure the hotness or coldness of a substance.
7. Pressure is measured by the SI unit _____ and the common units _____ of _____.
8. According to Boyle's law as the pressure increases the volume _____.
9. According to Charles' law as the volume increases the temperature _____.
10. A gas which obeys all of the gas laws is called an _____ gas.
11. No gases are truly ideal but most can be considered _____ under ordinary conditions.
12. At high pressure and or low temperature gases must be treated as _____ gases.
13. According to the kinetic molecular theory gases consist of particles in _____ motion.
14. According to kinetic molecular theory gas particles do not _____ with each other and have _____ volume.
15. According to kinetic molecular theory gas pressure results from particle _____ with the container walls.
16. According to kinetic molecular theory the temperature of a gas measures the average _____ of the gas particles.
17. Liquids have _____ volume but not a _____ shape.
18. Liquids take the _____ of their container but do not completely _____.
19. Liquids have _____ density measured in _____.
20. A liquid to gas change in state is called _____.
21. Vaporization in an open container is called _____.
22. Condensation is a _____ to _____ change in state.
23. Vaporization in a closed container continues until _____ is reached between condensation and vaporization.
24. Equilibrium can be either _____ or _____.
25. In _____ equilibrium nothing is actually happening.
26. In _____ equilibrium the changes are equal and opposite.
27. The pressure exerted by a vapor above its liquid at equilibrium is called the _____.
28. A _____ is the gaseous state of a substance which normally is not gaseous.
29. A substance with a high vapor pressure at room temperature is said to be _____.
30. A substance with a low vapor pressure at room temperature is said to be _____.
31. A liquid boils when the vapor pressure equals the _____ pressure.

Unit Six

32. When vapor pressure of a liquid = 1 atmosphere the liquid is at its _____.
33. When the external pressure is greater than 1 atm the boiling point will be _____ than the normal boiling point.
34. When the external pressure is less than 1 atm the boiling point will be _____ than the normal boiling point.
35. Solids have a _____ shape and a _____ volume.
36. Solids and liquids are not compressible as _____ are.
37. The temperature and pressure at which solid, liquid and gas states of a substance are at equilibrium is called the _____ point.
38. The temperature beyond which a substance in the gaseous state cannot be liquified regardless of the pressure is called the _____ temperature.
39. The pressure required to liquify a gas at the critical temperature is called the _____ pressure.
40. The state of substance beyond its critical point is the _____ fluid state.
41. One unusual property of water is that its density is maximum at a temperature _____ its freezing point namely _____ °C.
42. The density of ice is _____ than the density of liquid water.
43. Under extremely _____ pressure ice melts.
44. The properties of liquids are determined by _____ forces.
45. _____ forces act between polar molecules and _____ in strength as the temperature increases.
46. _____ is a special type of dipole—dipole force.
47. Hydrogen bonding occurs between H bonded to an O, N or F in one molecule and an _____, _____ or _____ in another molecule.
48. Hydrogen bonding has the _____ strength of all intermolecular forces and its strength is ~ _____ % that of a single covalent bond.
49. London dispersion forces are the only forces acting between _____ molecules.
50. Individual dispersion forces are weak but because of their large numbers collectively dispersion forces are the _____ important in determining properties except for those substances which experience _____ bonding.
51. The greater the intermolecular forces the _____ the melting point.
52. The greater the intermolecular forces the _____ the boiling point.
53. The greater the intermolecular forces the _____ the vapor pressure.
54. The unbalanced forces acting on surface molecules creates _____.
55. The greater the intermolecular forces the _____ the surface tension.
56. _____ is the attractive force between like molecules.
57. _____ is the attractive force between unlike molecules.
58. If cohesion is _____ than adhesion a liquid will wet the surface.
59. If cohesion is _____ than adhesion a liquid will not wet the surface.

INTRODUCTION TO CHEMISTRY

Review Problems

1. Perform the following conversions.
   a) 122°C to K
   b) 31.4°C to K
   c) −7.35°C to K
   d) 3.8×10^2 torr to atm
   e) 1.63×10^4 mm Hg to torr
   f) 7.60mm Hg to atm
   g) 3.0atm to torr
   h) 0.500 atm to mm Hg
   i) 10.0 Pa to atm

2. A gas occupies a 10.0L vessel at P = 4.00 atm. It is allowed to expand into a 20.0L vessel, at constant n and T. What is the new P?

3. A gas occupies a 40.0mL flask at $P = 3.60 \times 10^2$ torr. The pressure is increased to 1.44×10^3 torr, at constant n and T. What is the new V?

4. A gas occupies a 6.0L vessel at $T = 27°C$. It is allowed to expand into an 18.0L vessel, at constant n and P. What is the new T in °C?

5. A gas has P = 2.00 atm, V = 0.400L, T = −73°C. How many moles of gas are present?

6. A gas has $P = 3.80 \times 10^2$ torr, T = 127°C, n = 0.10 mol. What is the V in mL?

7. A gas has P = 76.0 mm Hg, V = 300.0mL, n = 0.0030 mol. What is the T in °C?

8. Gases A and B are mixed in a vessel so that P_A = 2.0atm, P_B = 4.0atm. What is P_{total}?

9. 0.20 mol of gas A is mixes with 0.30 mol of gas B. The total pressure = 10.0 atm. Determine X_A, P_A, X_B, P_B.

Answers to Review Problems

1.
   - a) 395K
   - b) 304.6K
   - c) 265.80K
   - d) 0.50 atm
   - e) 1.63×10^4 torr
   - f) 0.0100 atm
   - g) 2.3×10^3 torr
   - h) 3.80×10^2 mm Hg
   - i) 9.90×10^{-5} atm

2. 2.00 atm

3. 10.0 mL

4. 627°C

5. 0.0489 mol

6. 6.6×10^3 mL

7. -1.5×10^2 °C

8. 6.0 atm

9. $X_A = 0.40$ $X_B = 0.60$
   $P_A = 4.0$ atm $P_B = 6.0$ atm

Unit Six

7 Unit Seven

Solutions and Solution Stoichiometry

Unit Seven

I. **Solution**

  A. **Solutions consist of two components: solute and solvent**

  solute vs solvent

  B. **Types of solutions: The solute can be (g), (l), or (s) and the solvent can be (g), (l), (s)**

  solute in solvent   solute in solvent   solute in solvent

  C. **In chemistry the most important type of solutions are the:**

  D. **Steps in the dissolution of a solid in a liquid**
   1. Separation of the solute particles — The crystal lattice energy (CLE)

   2. Separation of the solvent particles — The solvent stripping energy (SSE)

   3. Solvation of the solute particles — The molar solvation energy (MHE)

Unit Seven

Hydration

E. **Energetics of the overall solution process for solid in liquid**

II. Solubility
A. **Factors which affect solubility**
   1. Nature of the solute and solvent

   2. Temperature effects
      a) For solid in liquid

      b) For gas in liquid

         Thermal pollution

   3. Pressure effects
      a) For solid in liquid

INTRODUCTION TO CHEMISTRY

b) For gas in liquid—Henry's Law

4. Factors which affect the rate of dissolution of a (s) in (*l*)
   a) Increasing the temperature

   b) Stirring

   c) Increasing the surface area

B. Equilibrium and solutions
   1. Saturated solutions

   2. Unsaturated solutions

   3. Supersaturated solutions

Unit Seven

C. For liquid in liquid solutions the extent of mixing ranges from

completely miscible   to   completely immiscible

III. Concentration of solutions

A. Concentration of solutions can be measured by

1. molarity

2. molality

3. $\%\dfrac{m}{m}$

4. mole fraction

B. Calculations involving molarity

1. Example
A solution has volume = 100.0 mL and contains 0.400 mol of solute. What is the molarity of the solution?
First convert mL to L: $100.0\,\text{mL}\left(\dfrac{1\,\text{L}}{10^3\,\text{mL}}\right) = 0.1000\,\text{L}$

Use definition of molarity $M = \dfrac{\#\text{ mol solute}}{\#\text{ L solution}} = \dfrac{0.400\,\text{mol}}{0.1000\,\text{L}} = 4.00\,\text{mol/L}$

2. More examples

Unit Seven

3. Example:
   A solution of volume = 300.0 mL has M = 2.00 mol/L. How many moles of solute are present?
   By dimensional analysis: Treat the given molarity as a conversion factor.

   $$300.0\,\text{mL}\left(\frac{1\,\text{L}}{10^3\,\text{mL}}\right)\left(\frac{2.00\text{ mol solute}}{1\text{ L soln}}\right) = 0.600\text{ mol solute}$$

   By formula application: mol A = (L of A)(M)

   mol A = (0.3000 L)(2.00 mol/L) = 0.600 mol solute

4. More examples

5. Example:
   What volume of a 0.100M solution contains 0.0200 mol of solute:

   By dimensional analysis:

   $$0.0200 \text{ mol}\left(\frac{1L}{0.100 \text{ mol}}\right)\left(\frac{10^3 \text{mL}}{1L}\right) = 2.00 \times 10^2 \text{mL solution}$$

   By formula application:

   L of A = mol A/M =

   $$\frac{0.0200 \text{ mol}}{0.100 \text{ mol/L}} = 0.200L \left(\frac{10^3 \text{mL}}{1L}\right) = 2.00 \times 10^2 \text{mL solution}$$

6. More examples

C. Calculations of molality

1. Example:
   A solution consists of 2.00 mol of solute and 200.0g of solvent. What is its molality?
   First convert g of solvent to Kg

   $$(200.0 \text{g solvent})\left(\frac{1 \text{kg}}{10^3 \text{g}}\right) = 0.2000 \text{kg solvent}$$

   Use the definition of molality = $\frac{2.00 \text{ mol}}{0.2000 \text{kg}}$ = 10.0 mol/kg

2. More examples

Unit Seven

D. **Calculations of %$\frac{m}{m}$**

   1. Example:
      A solution has mass = 40.0g and contains 10.0g of solute. What is its %$\frac{m}{m}$?

      Use the definition of %$\frac{m}{m}$ = $\frac{10.0g}{40.0g} \times 100$ = 25.0%

   2. More examples

E. **Calculations of mole fraction**

   1. Example:
      A solution contains 0.100 mol of solute and 0.900 mol of solvent. What is the mole fraction of solute?

      Use the definition of X_{solute} = $\frac{0.100 \text{ mol}}{0.100 + 0.900}$ = 0.100

   2. More examples

F. **Dilution Problems**
   Dilution formula: $M_{conc}V_{conc} = M_{dil}V_{dil}$
   where
   M_{conc} =
   V_{conc} =
   M_{dil} =
   V_{dil} =

   1. Example
      What volume of 10.0M HCl(aq) is needed to make 200.0mL of 4.00M HCl(aq)? How is the final solution prepared?
      M_{conc} = 10.0M, V_{conc} = ?, M_{dil} = 4.00M, V_{dil} = 200.0mL
      (10.0M)V_{conc} = (4.00M)(200.0mL)

INTRODUCTION TO CHEMISTRY

$$V_{conc} = \frac{(4.00M)(200.0mL)}{10.0M}$$

$V_{conc} = 80.0mL$

In making this solution remember AAA (Always Add Acid to water) when diluting concentrated acids.
Add 80.0mL of the 10.0M HCl(aq) to ~100mL H_2O. Then add sufficient water to the now diluted acid to have 200.0mL.

2. Other examples

IV. Solution Stoichiometry

A. Acid-Base titration experimental set-up

Unit Seven

B. **Acid-Base titration calculations**
1. Example:
   Consider the titration of 10.0mL of HCl(aq) with 0.200M NaOH(aq). The initial buret reading is 2.40mL. At the end-point the final buret reading is 17.40mL. Determine the molarity of the HCl(aq).
   The reaction is: $HCl(aq) + NaOH(aq) \rightarrow NaCl(aq) + H_2O(l)$

2. By dimensional analysis:
   Convert mL base to L:
   17.40
   −2.40
   $15.00 \text{mL NaOH} \left(\dfrac{1 \text{L}}{10^3 \text{mL}} \right) = 1.500 \times 10^{-2} \text{L NaOH}$

   Convert L base to mol base using the molarity as the conversion factor

   $1.500 \times 10^{-2} \text{L NaOH} \left(\dfrac{0.200 \text{ mol NaOH}}{1 \text{L NaOH}} \right) = 3.00 \times 10^{-3} \text{mol NaOH}$

   Convert mol base to mol acid using the balanced reaction.

   $(3.00 \times 10^{-3} \text{mol NaOH}) \left(\dfrac{1 \text{ mol HCl}}{1 \text{ mol NaOH}} \right) = 3.00 \times 10^{-3} \text{mol HCl}$

   Calculate molarity of acid.

   $M_{HCl} = \dfrac{3.00 \times 10^{-3} \text{mol HCl}}{10.0 \text{mL HCl} \left(\dfrac{1 \text{L}}{10^3 \text{mL}} \right)} = 0.300 \text{ mol/L}$

3. By formula application:
   Use the formula: $\dfrac{1}{a}(M_A V_A) = \dfrac{1}{b}(M_B V_B)$

   $\dfrac{1}{1}(M_A \cdot 10.0 \text{mL}) = \dfrac{1}{1}(0.200 \text{ mol/L} \cdot 15.00 \text{mL})$

   $M_a = \left(\dfrac{15.00}{10.0} \right)(0.200 \text{ mol/L}) = 0.300 \text{ mol/L}$

INTRODUCTION TO CHEMISTRY

C. More acid-base titration problems

V. Colligative properties of solutions
   A. Colligative properties
      1. Freezing point depression
      2. Boiling point elevation
      3. Vapor pressure lowering
      4. Osmotic pressure
         a) Osmosis
         b) semi-permeable membrane
         c) iso-osmotic
         d) hypo-osmotic
         e) hyper-osmotic

      5. Reverse osmosis

VI. Colloidal suspensions

    A.   Experimentally colloidal suspensions can be distinguished from solutions by the Tyndall effect
Tyndall effect

    B.   The suspended particles are stabilized by attracting a layer of ions around it.

    C.   These suspended particles can be coagulated by heating, application of an electric field, or the introduction of ions into the suspension

Summary of Calculations

Flowchart for dimensional analysis

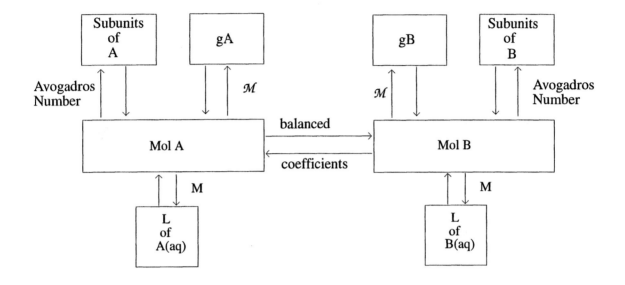

Formulas for the mole

$$\text{mol A} = \frac{gA}{\mathcal{M}_A}, \qquad \text{mol A} = \frac{\text{sub-units of } A}{\text{Avogadros \#}}, \qquad \text{mol A} = M_A V_A$$

Mole-to-mole relationship in a chemical reaction

$$\frac{1}{a}(\text{mol } A) = \frac{1}{b}(\text{mol } B)$$

$$\text{Molarity} = \frac{\text{mol solute}}{\text{L solution}}$$

$$\%\frac{m}{m} = \frac{\text{g solute}}{100\text{g solution}} \times 100$$

$$\text{mole fraction } A = \frac{\text{mol } A}{\text{total mols}}$$

$$\text{molality} = \frac{\text{mol solute}}{\text{kg solvent}}$$

$$M_{\text{conc}} V_{\text{conc}} = M_{\text{dil}} V_{\text{dil}}$$

Things to know

Definitions
1. solution
2. solute
3. solvent
4. solvation
5. hydration
6. solubility
7. Henry's Law
8. saturated solution
9. unsaturated solution
10. supersaturated solution
11. completely miscible
12. completely immiscible
13. molarity
14. molality
15. % $\frac{m}{m}$
16. mole fraction
17. colligative properties
18. end-point
19. colloidal suspension
20. Tyndall effect
21. osmosis
22. osmotic pressure
23. semi-permeable membrane
24. iso-osmotic
25. hypo-osmotic
26. hyper-osmotic
27. reverse osmosis

Concepts

1. Know the 9 types of solutions.
2. Know the 3 steps in the solution process for a solid in liquid.
3. Know the energetics of the steps in the solution process for a solid in liquid and for the overall process.
4. Know what "like dissolves like" means.

5. Know the effect of temperature on solid in liquid solutions.
6. Know the effect of temperature on gas in liquid solutions.
7. Know the effect of pressure on solid in liquid solutions.
8. Know the effect of pressure on gas in liquid solutions.
9. Know how thermal pollution originates.
10. Know Henry's law.
11. Know the factors which affect the rate of dissolution.
12. Be able to calculate molarity given moles of solute and volume of solution.
13. Be able to calculate moles of solute given molarity of solution and volume of solution.
14. Be able to calculate volume of solution given molarity of solution and moles of solute.
15. Be able to calculate molality from moles of solute and kg of solvent.
16. Be able to calculate $\%\frac{m}{m}$ from grams of solute and grams of solution.
17. Be able to calculate mole fraction given moles of solute and moles of solvent.
18. Know the experimental set-up for an acid-base titration.
19. Be able to calculate molarity of the acid in an acid-base titration.
20. Know how the freezing point of a solution compares to that of the pure solvent.
21. Know how the boiling point of a solution compares to that of the pure solvent.
22. Know how the vapor pressure of a solution compares to that of the pure solvent.
23. Know how suspended colloidal particles are stabilized.
24. Know how suspended colloidal particles can be coagulated.
25. Be able to explain the Tyndall effect and how it distinguishes solutions and colloidal suspensions.

Unit Seven

Review Questions

1. Another name for a solution is a _____ mixture.
2. The substance dissolved in a solution is the _____.
3. The dissolving medium in a solution is the _____.
4. Based on physical states how many solution types are there? _____
5. The surrounding of solute particles by solvent particles is called _____.
6. The surrounding of solute particles by water particles is called _____.
7. For a (s) in (l) type solution the energy change may be _____ or _____.
8. The grams of solute which can be dissolved in 100g of solvent is called _____.
9. Polar and ionic solutes are soluble in _____ solvents.
10. Non-polar solutes are soluble in _____ solvents.
11. For (s) in (l) solutions an increase in temperature will cause solubility to _____ or _____.
12. For (g) in (l) solutions an increase in temperature will cause solubility to _____.
13. For (s) in (l) solutions an increase in pressure will cause solubility to _____.
14. For (g) in (l) solutions an increase in pressure will cause solubility to _____.
15. Increasing the temperature of a (s) in (l) solution will cause the rate of dissolution to _____.
16. Stirring a (s) in (l) solution will cause the rate of dissolution to _____.
17. Increasing the surface area of a solid solute will _____ the rate of dissolution.
18. A solution in which dissolution and crystallization are at equilibrium is called a _____ solution.
19. A solution which contains the maximum amount of solute at a given temperature is called a _____ solution.
20. A solution which contains less than the maximum amount of solute at a given temperature is called a _____ solution.
21. A solution which contains more than the maximum amount of solute at a given temperature is called a _____ solution.
22. A supersaturated solution can be made by allowing a _____ saturated solution to _____.
23. When two liquids mix in all proportions they are siad to be completely _____.
24. When two liquids mix in no proportions they are said to be completely _____.
25. In an acid-base titration the point at which the acid is just neutralized is called the _____.
26. To visually detect the end-point in an acid-base titration a chemical dye called an _____ is added to the acid solution.
27. The most commonly used indicator is _____.
28. In an acid solution phenolphthalein is _____ while in base solution it is _____.

Unit Seven

29. When diluting concentrated acids always add _____ to _____.
30. Properties of solutions which depend only on the number of dissolved particles are called _____ properties.
31. Compared to the pure solvent a solution has a _____ freezing point.
32. Compared to the pure solvent a solution has a _____ boiling point.
33. Compared to the pure solvent a solution has a _____ vapor pressure.
34. The flow of solvent across a semi-permeable membrane from low solute to high solute concentration is called _____.
35. A _____ membrane allows only the passage of solvent particles.
36. The pressure required to prevent osmosis into a solution from the pure solvent is called _____ pressure.
37. Two solutions with the same osmotic pressure are said to be _____ or _____.
38. A solution with a lower osmotic pressure than some reference solution is said to be _____ or _____.
39. A solution with a higher osmotic pressure than some reference solution is said to be _____ or _____.
40. _____ osmosis occurs if a pressure greater than the osmotic pressure is applied to a solution.
41. Reverse osmosis is one method used in the _____ of water.
42. A suspension of particles with diameters between 1 and 1000nm in a dispersing medium is called a _____ suspension.
43. A colloidal suspension can be distinguished from a solution by the _____ effect.
44. The Tyndall effect occurs when light is _____ off suspended particles.
45. Suspended particles are stabilized by the _____ repulsion due to a layer of _____ around each particle.
46. Suspended particles can be coagulated by _____, application of an _____ field or the introduction of _____ into the suspension.

Review Problems

1. Determine the molarity of each solution.
   a) 0.60 mol solute in 4.00L solution
   b) 0.10 mol solute in 400.0mL solution
   c) 3.00 mol solute in 900.0mL solution
   d) 2.00×10^{-3} mol solute in 40.0mL solution

2. Determine the moles of solute present in each solution.
   a) $V = 200.0$mL, $M = 1.60$ mol/L
   b) $V = 1.50$L, $M = 0.400$ mol/L
   c) $V = 10.0$mL, $M = 12.0$ mol/L
   d) $V = 20.0$L, $M = 0.060$ mol/L

3. Determine the volume in mL of each solution needed to have the indicated moles of solute.
   a) $M = 0.300$ mol/L, 0.060 mol
   b) $M = 1.20$ mol/L, 3.60 mol
   c) $M = 4.00$ mol/L, 2.0×10^{-3} mol
   d) $M = 0.100$ mol/L, 4.0 mol

4. Determine the molality of each solution.
   a) 0.20 mol in 4.00kg solvent
   b) 4.00 mol solute in 0.200kg solvent
   c) 10.0 mol in 1.50×10^4 g solvent
   d) 0.600 mol in 30.0g solvent

5. Determine the $\%\frac{m}{m}$ of solute in each solution.
   a) 30.0g solute in 100.0g solution
   b) 4.0g solute in 80.0g solution
   c) 2.4×10^2 g solute in 7.2×10^3 g solution
   d) 3.0mg solute in 30.0mg solution

6. Determine the mole fraction of solute in each solution.
   a) 0.200 mol solute in 0.800 mol solvent
   b) 1.00 mol solute in 3.00 mol solvent
   c) 0.0200 mol solute in 1.88 mol solvent
   d) 0.400 mol solute in 0.500 mol solvent

7. Determine the mL of stock needed to perform the indicated dilution.
   a) 100.0mL of 2.00M from 10.0M stock
   b) 500.0mL of 0.100M from 4.00M stock
   c) 3.00L of 0.600M from 12.0M stock
   d) 20.0mL of 0.200M from 14.0 stock

8. Determine the molarity of the acid from each titration experiment.

a) 10.0mL of HNO_3 is titrated by KOH to the end-point. The molarity of the KOH = 2.00M, initial buret volume = 0.40mL, final buret volume = 15.40mL. The reaction is
$HNO_3(aq) + KOH(aq) \rightarrow KNO_3(aq) + H_2O(l)$.

b) 20.0mL of H_2SO_4 is titrated with 0.400M NaOH to the end-point. The initial buret volume = 4.00mL, final buret volume = 8.00mL. The reaction is
$H_2SO_4(aq) + 2NaOH(aq) \rightarrow Na_2SO_4(aq) + 2H_2O(l)$.

Answers to Review Problems

1.
   a) 0.15 mol/L
   b) 0.25 mol/L
   c) 3.33 mol/L
   d) 0.0500 mol/L

2.
   a) 0.320 mol
   b) 0.600 mol
   c) 0.120 mol
   d) 1.2 mol

3.
   a) 2.0×10^2 mL
   b) 3.00×10^3 mL
   c) 0.50 mL
   d) 4.0×10^4 mL

4.
   a) 0.050 mol/kg
   b) 20.0 m
   c) 0.667 mol/kg
   d) 20.0 m

5.
   a) $30.0\% \frac{m}{m}$
   b) $5.0\% \frac{m}{m}$
   c) $3.3\% \frac{m}{m}$
   d) $1.0 \times 10^1 \% \frac{m}{m}$

6.
   a) 0.200
   b) 0.250
   c) 0.0100
   d) 0.444

7.
   a) 20.0 mL
   b) 12.5 mL
   c) 1.50×10^2 mL
   d) 0.286 mL

8.
   a) 3.00 M
   b) 0.0400 M

8 Unit Eight

Acids and Bases
Kinetic
Equilibrium
Thermodynamic

Unit Eight

I. **Acids and Bases**

    A. **Arrhenius definition of an acid**

        Examples of Arrhenius acids

    B. **Arrhenius definition of a base**

        Examples of Arrhenius bases

    C. **The concentration of acid and base solutions is measured by the pH scale**

    D. **The pH scale**

    E. **Each difference of one pH unit corresponds to a factor of 10 in solution strength**

        1. Example:
   Solution A has pH = 3.0, solution B has pH = 5.0. Compare the acid strength of the two solutions.
   A difference of 2 pH units means solution A is $10 \times 10 = 100$ times stronger than B.

        2. More examples

F. The Arrhenius acid-base definitions are the most restrictive.

G. A more general set of definitions is due to Bronsted-Lowry
   1. Bronsted-Lowry acid

   2. Bronsted-Lowry base

H. Bronsted-Lowry acids and bases come in pairs called conjugate acid-base pairs.
   1. Conjugate base

   2. Conjugate acid

I. Example of a Bronsted-Lowry acid-base reaction
$$HCl(aq) + OH^-(aq) \rightarrow H_2O(l) + Cl^-(aq)$$
   Acid$_1$ Base$_2$ Conjugate acid$_2$ Conjugate base$_1$

J. pH changes during the titration of strong acid with strong base

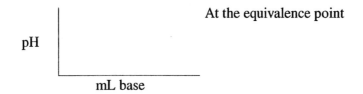

K. pH changes curing the titration of a weak acid with a stong base

```
         |                At the equivalence point
    pH   |
         |_____
             mL base
```

II. Thermodynamics

- **A. System**
- **B. Surroundings**
- **C. Types of systems**
    1. Open
    2. Closed
    3. Isolated
- **D. 1st Law of Thermodynamics—Conservation of energy**

$$\Delta E = q + W$$

ΔE
q
W
Examples of 1st Law problems

Unit Eight

E. The heat change at constant p, q_p, has special significance and is called the enthalpy H. Only ΔH can be determined for a reaction. Whether ΔH is positive or negative decides if a reaction is endothermic or exothermic.

1. $\Delta H > 0$ the reaction is

2. $\Delta H < 0$ the reaction is

F. Other properties of ΔH

1. Reversing a reaction changes the sign of ΔH

2. ΔH is extensive

G. Entropy

H. Change in entropy = ΔS

I. Sign of ΔS for various physical and chemical processes

J. Spontaneity

K. Predicting spontaneity

1. ΔH alone is not sufficient

INTRODUCTION TO CHEMISTRY 305

2. ΔH and ΔS together is sufficient
3. $\Delta G = \Delta H - T\Delta S$
4. ΔG = change in Gibb's free energy
5. The sign of ΔG depends on the sign of ΔH and ΔS

sign of ΔH	sign of ΔS	sign of ΔG	spontaneity

III. Kinetics

A. Factors which affect the rate of a reaction
1. Concentration of reactants

2. Temperature

3. Presence of a catalyst

B. Collision theory of reaction rates
1. Chemical reactions occur when the reactants collide if the collision has
    a) sufficient energy

    b) proper orientation

C. Reaction profiles
1. For endothermic reactions

H │
  │
  └──────────
    reaction coordinate

2. For exothermic reactions

H │
  │
  └──────────
    reaction coordinate

3. Activation energy
4. Activated complex or transition state

IV. Chemical Equilibrium

A. Condition for equilibrium:

rate of forward reaction = rate of reverse reaction

B. Symbolism

C. The Equilibrium Constant: K_c

1. Only (aq) and (g) species are included
2. Pure (l) and (s) are not included
3. For the general reaction: Assume all species are (aq) or (g)

$$aA + bB \rightarrow cC + dD$$

$$K_c =$$

4. specific examples

D. LeChatelier's Principle

E. Types of Stress
1. Changes in concentration
   Examples

2. Changes in temperature
   Examples

3. Changes in volume

Things to Know

Terms
1. Arrhenius acid
2. Arrhenius base
3. pH
4. Bronsted-Lowry acid
5. Bronsted-Lowry base
6. conjugate base
7. conjugate acid
8. equivalence point
9. system
10. surroundings
11. open system
12. closed system
13. isolated system
14. q
15. w
16. ΔE
17. enthalpy
18. endothermic
19. exothermic
20. entropy
21. spontaneity
22. ΔG
23. catalyst
24. effective collisions
25. activation energy
26. transition state
27. chemical equilibrium
28. LeChatelier's principle
29. 1st Law of thermodynamics
30. 2nd Law of thermodynamics

Concepts

1. Know the typical range of the pH scale
2. Be able to decide if a solution is acidic or basic from its pH value
3. From a comparison of pH values be able to compare acid or base strength of solutions
4. Be able to recognize conjugate acid-base pairs in a reaction
5. Be able to draw and explain pH curves for strong base-strong acid and strong base-weak acid titrations

6. Be able to work 1st Law of thermodynamics problems
7. Understand the relation between ΔH and q
8. Know the mathematical properties of ΔH
9. Be able to determine the sign of ΔS for various physical and chemical processes
10. Know the Gibbs-Helmholtz equation
11. Know the relation between ΔG and spontaneity
12. Know the how the sign of ΔG depends on the sign of ΔH and ΔS
13. Know the various factors which affect reaction rates
14. Know the various factors for a collision to be effective
15. Be able to draw and label reaction profiles for endothermic and exothermic reactions
16. Know how a catalyst affects reaction rates by changing the reaction profile
17. Know the condition for chemical equilibrium
18. Know which species are and are not included in the equilibrium constant expression
19. Be able to write equilibrium constant expressions for reactions
20. Be able to use LeChatelier's principle to determine how a system in equilibrium shifts under various types of stress

Review Questions

1. An Arrhenius _____ produces $H^+(aq)$ when dissolved in water.
2. An Arrhenius _____ produces $OH^-(aq)$ when dissolved in water.
3. The concentration of acid and base solutions is measured by the _____ scale.
4. The pH of a neutral solution is exactly _____.
5. The pH of a basic solution is _____ than 7.
6. The pH of an acidic solution is _____ than 7.
7. For most solutions the pH scale runs from _____ to _____.
8. Each difference of one pH unit corresponds to a factor of _____ in solution strength.
9. A Bronsted-Lowry _____ is a proton donor.
10. A Bronsted-Lowry _____ is a proton acceptor.
11. Bronsted-Lowry acid and bases come in pairs called _____ acid-base pairs.
12. A conjugate _____ results when an acid loses an H^+.
13. A conjugate _____ results when a base gains on H^+.
14. In the titration of a strong acid with a strong base the pH at the equivalence point is _____.
15. In the titration of a weak acid with a strong base the pH at the equivalence point is _____ than 7.
16. The part of the universe being studied is the _____.
17. The rest of the universe outside of the system is the _____.
18. A _____ system exchanges both mass and energy with the surroundings.
19. A _____ system exchanges energy but not mass with the surroundings.
20. A _____ system exchanges neither energy nor mass with the surroundings.
21. According to the first law of thermodynamics ΔE = _____ + _____.
22. ΔE is the change in _____ energy of the system.
23. q is the change in _____ energy of the system.
24. W is the _____ done on or by the system.
25. The sign of q is _____ if the heat energy of a system increases.
26. The sign of W is _____ if work is done on a system.
27. The heat change at constant pressure is called the _____ of a system.
28. The heat change at constant pressure is symbolized by _____.
29. The change in entropy of a system is symbolized by _____.
30. The sign of ΔH is _____ if a reaction is exothermic.
31. The sign of ΔH is _____ if a reaction is endothermic.
32. If a reaction is reverse the sign of ΔH _____.
33. If a reaction is multiplied by n the value of ΔH is _____ by n.
34. According to Hess' Law ΔH values, like reactions, are _____.

35. The disorder of a system is measured by its _____.
36. The symbol for change in entropy of a system is _____.
37. If the number of moles of gas of a system increases the sign of ΔS is _____.
38. If the temperature of a system increases the sign of ΔS is _____.
39. For the solution process the sign of ΔS is _____.
40. For $(s) \rightarrow (l)$, $(l) \rightarrow (g)$, or $(s) \rightarrow (g)$ changes in state the sign of ΔS is _____.
41. For $(g) \rightarrow (l)$, $(l) \rightarrow (s)$, or $(g) \rightarrow (s)$ changes in state the sign of ΔS is _____.
42. A process is _____ if it occurs naturally, on its own, in the forward direction.
43. Most, but not all, _____ reactions are spontaneous.
44. The symbol for the change in Gibb's free energy is _____.
45. $\Delta G =$ _____ $-$ _____.
46. If ΔH is negative and ΔS is positive the sign of ΔG is _____ and the process is _____.
47. If ΔH is positive and ΔS is negative the sign of ΔG is _____ and the process is _____.
48. If ΔH is negative and ΔS is negative the sign of ΔG depends on the _____.
    At low temperature the sign of ΔG is _____ and the process is _____.

    At high temperature the sign of ΔG is _____ and the process is _____.

49. If ΔH is positive and ΔS is positive the sign of ΔG depends on the _____.
    At low temperature the sign of ΔG is _____ and the process is _____.

    At high temperature the sign of ΔG is _____ and the process is _____.

50. As the reactant concentrations increase the reaction rate will _____.
51. As the temperature of a reaction increases the reaction rate will _____.
52. In the presence of a catalyst the reaction rate will _____.
53. For a collision to produce a reaction the collision must have the proper _____ and sufficient _____.
54. The energy required to start a reaction is called the _____ energy.
55. In general the activation energy of an exothermic reaction is _____ than that for an endothermic reaction.
56. At the peak of a reaction profile the reactants and products exist in the form of an _____ complex or _____ state.

57. For a reaction to be in equilibrium the rate of the forward and reverse reactions must be _____.
58. A reaction in equilibrium is symbolized by _____.
59. In writing an equilibrium constant only species in the _____ or _____ state are to be included.
60. In writing an equilibrium constant the balanced coefficients appear as _____.
61. In writing an equilibrium constant the numerator is composed of _____ and the denominator is composed of _____.
62. When a system in equilibrium is stressed the system will shift in such a way as to _____ the effect of the stress. This is _____ Principle.
63. If the reactant concentrations are increased in a system in equilibrium the system will shift to the _____.
64. If the product concentrations are increased in a system in equilibrium the system will shift to the _____.
65. If heat energy is added to an endothermic reaction in equilibrium the system will shift to the _____.
66. If heat energy is added to an exothermic reaction in equilibrium the system will shift to the _____.

Review Problems

1. Based on the given pH identify each solution as acidic, basic or neutral.
   a) pH = 2.0
   b) pH = 8.0
   c) pH = 7.0
   d) pH = 0.00
   e) pH = 14.0

2. Based on the given pH values of solution A and B compare their relative strengths as acids or bases.
   a) A = pH = 4.0
      B = pH = 5.0
   b) A = pH = 2.0
      B = pH = 6.0
   c) A = pH = 8.0
      B = pH = 11.0
   d) A = pH = 14.0
      B = pH = 9.0

3. Determine the conjugate base of each.
   a) HNO_3
   b) $HClO_4$
   c) $HC_2H_3O_2$
   d) H_2O

4. Determine the conjugate acid of each.
   a) NH_3
   b) OH^-
   c) Br^-
   d) H_2O

5. Determine ΔE for each process. Show set-up.
   a) system loses 10J of heat energy and has 20J of work done on it
   b) system gains 18J of heat energy and does 12J of work on the surroundings
   c) system loses 21J of heat energy and does 14J of work on the surroundings

6. Determine the new ΔH value.
   a) reaction has $\Delta H = 42J$, the reaction is reversed, new $\Delta H = ?$
   b) reaction has $\Delta H = 36J$, the reaction is doubled, new $\Delta H = ?$
   c) reaction has $\Delta H = 66J$, the reaction is reduced by $\frac{1}{3}$, new $\Delta H = ?$
   d) reaction A has $\Delta H = 22J$, reaction B has $\Delta H = 33J$, for reaction $A + B$, $\Delta H = ?$

7. Determine K_c for each reaction.
   a) $2C_4H_{10}(g) + 13O_2(g) \rightleftharpoons 8CO_2(g) + 10H_2O(g)$
   b) $2NaCl(aq) + 2H_2O(l) \rightleftharpoons 2NaOH(aq) + H_2(g) + Cl_2(g)$
   c) $BaSO_4(s) \rightleftharpoons Ba^{2+}(aq) + SO_4^{2-}(aq)$

8. Determine the direction of shift in the following equilibrium given the indicated stresses. $N_2(g) + 3H_2(g) \rightleftharpoons 2NH_3(g) \quad \Delta H = -92\,kJ$
   a) The concentration of $H_2(g)$ is increased.
   b) The concentration of $NH_3(g)$ is decreased.
   c) The concentration of $N_2(g)$ is decreased.
   d) The temperature is increased.
   e) The temperature is decreased.
   f) The volume is decreased.

Solutions to Review Problems

1.   a) acidic   c) neutral   e) basic
       b) basic   d) acidic

2.   a) A is 10 times more acidic than B
       b) A is 10,000 times more acidic than B
       c) B is 1000 times more basic than A
       s) B is 100,000 times more basic than A

3.   a) NO_3^{1-}
       b) ClO_4^{1-}
       c) $C_2H_3O_2^{1-}$
       d) OH^-

4.   a) NH_4^{1+}
       b) H_2O
       c) HBr
       d) H_3O^{1+}

5.   a) $\Delta E = -10J + 20J = 10J$
       b) $\Delta E = 18J - 12J = 6J$
       c) $\Delta E = -21J - 14J = -35J$

6.   a) $\Delta H = -42J$
       b) $\Delta H = 72J$
       c) $\Delta H = 22J$
       d) $\Delta H = 55J$

7.   a) $K_c = \dfrac{[CO_2]^8[H_2O]^{10}}{[C_4H_{10}]^2[O_2]^{13}}$

       b) $K_c = \dfrac{[NaOH]^2[H_2][Cl_2]}{[NaCl]^2}$

       c) $K_c = [Ba^{2+}]^8[SO_4^{2-}]$

8.   a) →
       b) →
       c) ←
       d) ←
       e) →
       f) →

9 Unit Nine

Nuclear Chemistry

Unit Nine

I. Nuclear Structure

A. The nucleus

B. Properties of the subatomic particles

Particle	Symbol	Charge	Mass (g)	Mass (amu)
proton				
neutron				
electron				

II. Nuclear symbolism of the elements: $^A_Z X$

III. Nuclear stability

IV. The discovery of radioactivity

A. 3 common types of radioactive decay

1. Alpha particle production
   a) Properties of the α particle

   b) Examples

2. Beta particle production
    a) Properties of β particles

    b) Examples

3. Gamma ray production
    a) Properties of γ rays

    b) Examples

B. **Half-life of radioactive isotopes**

C. **Nuclear Equations**
    1. radioactive nucleus → new nucleus + radiation
    2. mass numbers and atomic numbers must balance
    3. Examples

D. **Characteristics of Radiation**
   1. Ionizing ability

   2. Penetrating ability

   3. Intensity

   4. Shielding

   5. Detection

   6. Measuring units
      a) Curie

      b) Rad

      c) Rem

7. Biological effects of radiation exposure
    a) Internal vs External exposure

    b) Background Radiation - no dose is safe

    c) Radiation sickness

E. **Medical Applications of Radioactive Isotopes**
    1. Diagnosis
        a) ^{131}I

        b) ^{99m}Tc

        c) ^{11}C - PET

Unit Nine

F. Therapy
 1. ^{60}Co

 2. ^{67}Ga

G. Producing radioactive isotopes by artificial transmutation
Examples

H. Other uses of radioactive isotopes
 1. Carbon dating

I. Nuclear Fission
 1. The fission of ^{235}U

Unit Nine

      2. Chain reactions

J.   **Nuclear Explosions**

K.   **Nuclear Reactors**

L.   **Nuclear Fusion**

Unit Nine

Things to Know

Terms
1. nucleus
2. proton
3. neutron
4. electron
5. atomic number
6. mass number
7. isotope
8. α type radioactivity
9. β type radioactivity
10. γ type radioactivity
11. half-life
12. Curie
13. rad
14. rem
15. roentgen
16. artificial transmutation
17. fission
18. chain reaction
19. critical mass
20. moderators
21. control rods
22. fusion
23. mass defect
24. nuclear binding energy
25. Geiger counter
26. RBE
27. LD_{50}
28. PET

Concepts

1. Know the symbol, charge and appropriate mass in amu for the proton, neutron, electron
2. From a nuclear symbol determine the number of protons and neutrons in the nucleus
3. Be able to explain why some nuclei are stable and others are not
4. Know properties of α, β, γ types of radioactivity
5. Be able to complete and balance nuclear equations
6. Know the ionizing properties of α, β, γ rays
7. Know the penetration ability of α, β, γ rays
8. Know how the intensity of a radioactive source varies with distance
9. Know what type of materials are needed to shield α, β, γ rays

INTRODUCTION TO CHEMISTRY

10. Know how radioactivity is detected
11. Know the various units used to measure radioactivity
12. Know the biological effects of both internal and external exposure to α, β, γ rays
13. Know the approximate value and effect of background radiation
14. Know the symptoms of radiation sickness
15. Know the diagnostic applications of ^{131}I, ^{99m}Tc, ^{11}C
16. Know the therapeutic applications of ^{60}Co, ^{67}Ga
17. Be able to describe how carbon dating works
18. Be able to describe the fission of ^{235}U
19. Know how a chain reaction occurs in ^{235}U
20. Know the result of an uncontrolled chain reaction
21. Be able to describe how a nuclear reactor works
22. Know the main drawbacks of nuclear power
23. Know the advantages and disadvantages of fusion over fission

Review Questions

1. The nucleus of an atom contains most of the _____ of the atom.
2. The diameter of the nucleus is about _____ the diameter of the atom.
3. The nucleus consists of _____ and _____.
4. The proton has a charge of _____ and a mass of _____ amu.
5. The neutron has a charge of _____ and a mass of _____ amu.
6. The number of protons in the nucleus is the _____ number.
7. The number of protons + neutrons in the nucleus is the _____ number.
8. In the nuclear symbol $^A_Z X$, A is the _____ number, Z is the _____ number.
9. The mass lost when a nucleus forms is the _____ .
10. The energy equivalent of the mass defect is the _____ .
11. The nuclear binding energy holds the _____ together.
12. When the attractive and repulsive forces within the nucleus do not balance the nucleus is _____ .
13. An unstable nucleus will naturally release energy in order to gain stability in a process called _____ .
14. In 1896 the scientist _____ accidentally discovered radioactivity.
15. The α particle has charge of _____ , mass of ~ _____ amu and is a _____ nucleus.
16. The β particle has charge of _____ , mass of ~ _____ amu and is an _____ .
17. The γ ray has charge of _____ , mass of _____ , and is a form of _____ energy.
18. The time required for 1/2 of a radioactive substance to decay is called the _____ .
19. Half lives may be very _____ or very _____ .
20. Nuclear equations must be _____ just like chemical equations.
21. In balancing nuclear equations the atomic numbers and mass numbers must be _____ on both sides of the equation.
22. The α particles have an ionizing ability about _____ times that of β and γ rays.
23. Listed in order of most penetrating to least penetrating: _____ > _____ > _____ .
24. The intensity of a radioactive source varies as the _____ of the distance from the source.
25. α particles can be shielded with _____ or _____ .
26. β particles can be shielded with _____ or _____ .
27. γ rays can be shielded with _____ or _____ .
28. The most common device used to detect radiation is the _____ .

29. One Curie = _____ disintegrations/second.
30. The Curie is too large for most medical uses and instead a more common unit for activity is the _____ or the _____ .
31. Rad stands for _____ _____ _____ .
32. The rad measures the amount of radiation absorbed by one _____ of tissue.
33. The most important biological measure of radiation is the _____ .
34. Rem stands for _____ _____ _____ .
35. Rem = _____ × _____ .
36. RBE stands for _____ _____ _____ ,
37. For α particles the RBE value is _____ .
38. For β particles and γ rays the RBE value is _____ .
39. Becuase of their low penetrating but high ionizing ability α particles are most dangerous when the source is _____ to the body.
40. Because of their high penetrating but low ionizing ability β particles and γ rays are most dangerous when the source is _____ to the body.
41. Because of natural and man-made sources the average person in the US receives about _____ millirems of background radiation annually.
42. Even very small doses of radiation can cause _____ and genetic _____ .
43. The lethal dose for 1/2 the population is symbolized by _____ .
44. The LD_{50} for humans is _____ rems.
45. If the exposure to radiation is 100 rems or higher the person will suffer _____ sickness which has the symptoms _____ , _____ , _____ _____ in white cell count.
46. If the exposure to radiation is 300 rems or higher the patient suffers _____ , _____ , _____ .
47. An exposure of _____ rem or higher will be fatal to all humans within a few weeks.
48. The two uses of radioisotopes in medicine are _____ and _____ .
49. For diagnosis the best radioisotopes are those which emit _____ and have _____ half-lives.
50. For diagnosis of thyroid malfunction small doses of _____ are used.
51. For scans of the brain, bone, lung, liver and other organs small doses of _____ are used.
52. PET stands for _____ _____ _____ .
53. A positron is the _____ of the electron. It has the same mass as the electron but a charge of _____ instead of −1.
54. PET is used to trace the _____ pathway in the brain and uses the radioisotope _____ .
55. For therapy sources with _____ half lives might be used.

INTRODUCTION TO CHEMISTRY

Unit Nine

56. For therapy internal sources are used which emit _____ or _____ .
57. For therapy external sources are used which emit _____ .
58. The two main external sources of radiation in the treatment of cancer are _____, _____ .
59. Those radioisotopes which are not naturally occurring may be produced through artificial _____ .
60. Artificial transmutation involves bombarding a stable non-radioactive atom with fast moving _____, _____, _____ .
61. The radioisotope used in carbon dating is _____ .
62. The fission of a nucleus produces nuclei with _____ mass numbers and releases large amounts of _____ .
63. The fissionable isotope of uranium is _____ .
64. The mass required to have a self-sustaining fission reaction is the _____ mass.
65. A self-sustaining fission reaction is called a _____ reaction.
66. An uncontrolled chain reaction produces a nuclear _____ .
67. The percentage of ^{235}U in naturally occuring uranium samples is _____ .
68. In nuclear weapons the percentage of ^{235}U must be _____ .
69. In nuclear reactors useful energy is produced by _____ chain reactions.
70. For nuclear reactors the percentage of ^{235}U must be _____ .
71. In nuclear reactors the energy released is used to convert water into _____ which is then used to produce _____ .
72. The combining of small nuclei into a large one is called nuclear _____ .
73. Extremely high _____ are required for nuclear fusion.
74. The energy released in fusion is _____ than that released in fission.
75. Unlike fission the waste products in fusion are not _____ .

INTRODUCTION TO CHEMISTRY

Addendum One
Chemical Nomenclature

Compounds can be classified into two broad naming groups: binary and ternary

Binary Compounds vs Ternary Compounds

Binary compounds can be broken down into three subgroups: Ionic, Covalent, Acids

Ionic compounds vs Covalent compounds vs Binary acids

Ternary compounds can be broken down into two subgroups: Ionic and Acids

Ternary ionic vs Ternary acids

Rules for naming binary ionic compounds
1. Name the cation: Its parent metal may be main group or transition

Main group vs transition

Example

2. Name the anion:

Examples

Addendum One

3. Combine the names: cation first, anion second
   Examples

Rules for naming binary covalent compounds
1. Indicate the number of atoms of each non-metal by the appropriate Greek prefix

Table of Greek Prefixes

Number equivalent	Prefix	Number equivalent	Prefix
1		6	
2		7	
3		8	
4		9	
5		10	

4. The prefix "mono-" is not used with the first element
   Examples

Rules for naming binary acids
1. The prefix is "hydro-"
2. The root name comes from the non-metal or polyatomic
3. The suffix is "-ic" followed by the word "acid".
4. Examples

Addendum One

Addendum One

Rules for naming ternary ionic compounds
1. Name the metal - it may be either main group or a transition metal
2. Name the polyatomic ion

List of Common Polyatomics

Ion	Name	Ion	Name	Ion	Name
NO_2^{1-}		CO_3^{2-}		BrO_3^{1-}	
NO_3^{1-}		PO_4^{3-}		BrO_2^{1-}	
SO_3^{2-}		MnO_4^-		BrO^{1-}	
SO_4^{2-}		HPO_4^{2-}		BrO_4^{1-}	
HSO_4^{1-}		$H_2PO_4^-$		CrO_4^{2-}	
ClO_3^{1-}		CN^-		$Cr_2O_7^{2-}$	
ClO_2^{1-}		OH^-			
ClO^{1-}		HCO_3^-			
ClO_4^{1-}		NH_4^+			

Patterns in the polyatomic ion names
1. Removing one oxygen from the -ate form gives the -ite form
2. Removing a second oxygen from the -ate form gives the hypo- -ite form
3. Add one oxygen to the -ate form gives the per- -ate form
4. Adding a hydrogen reduces the charge by +1 and adds the prefix hydrogen-
Examples of ternary ionic naming

Rules for naming ternary acids
1. Use no prefix except the one which comes with the polyatomic ion
2. If the polyatomic ion ends in -ate it changes to -ic
3. If the polyatomic ion ends in -ite it changes to -ous
4. Add the word acid at the end
Examples of naming ternary acids

INTRODUCTION TO CHEMISTRY

Addendum One

Going in reverse - From the name write the formula
Binary ionic
    Type I examples

    Type 2 examples

Binary acid examples

Ternary ionic examples

Ternary acid examples

Review Questions

1. Compounds can be classified into two naming groups: _____ and _____.
2. Binary compounds consist of _____ kinds of atoms.
3. Ternary compounds consist of _____ or more kinds of atoms.
4. Binary compounds can be _____ , _____ , or _____ .
5. Binary ionic consists of a _____ ion and a _____ ion.
6. Binary covalent consists of a _____ atom and a _____ atom.
7. Binary acids consist of _____ and a _____ atom.
8. Ternary compounds can be _____ or _____ .
9. Ternary ionic compounds consist of a _____ ion and a _____ ion.
10. Ternary acids consist of _____ and a _____ ion.
11. In naming main group cations the ion name is the same as the _____ name.
12. In naming transition metals the ion name is the metal name with the _____ indicated in _____ numerals in parenthesis after the name.
13. An anion in a binary ionic compound is named using the _____ of the element's name and adding the suffix _____ .
14. In naming binary covalent compounds the number of each type of atom is indicated by a _____ prefix.
15. The prefix _____ is not used with the first element.
16. In naming binary covalent compounds the first element uses its _____ name however the second element is named like the _____ in a binary ionic.
17. In naming binary acids the prefix _____ is used with the _____ of the non-metal name and the suffix _____ followed by the word _____ .
18. In naming ternary ionic compounds the metal is named either as a _____ or _____ element and the _____ ion is named second.
19. Removing one "O" from a polyatomic changes the "ate" ending to _____ .
20. Removing two "O" from a polyatomic changes the "ate" ending to _____ and adds the prefix _____ .
21. Adding one "O" to a polyatomic leaves the "ate" ending but adds the prefix _____ .
22. Adding a "H" to a polyatomic reduces the charge by _____ and adds the prefix _____ .
23. In naming ternary acids no prefix is used except one which comes with the _____ ion.
24. In naming ternary acids if the polyatomic ion ends in "ate" it is changed to _____ .
25. In naming ternary acids if the polyatomic ion ends in "ite" it is changed to _____ .
26. H and a polyatomic ion not containing oxygen is named as a _____ acid.

Review Problems

1. Name each binary ionic compound
   a) NaBr
   b) Al_2S_3
   c) K_2O
   d) CaS
   e) MgI_2
   f) $CuCl_2$
   g) Fe_2O_3
   h) $PbCl_4$
   i) Hg_2Cl_2

2. Name each binary covalent compound
   a) CO
   b) CO_2
   c) N_2O
   d) N_2O_4
   e) PCl_5
   f) Cl_2O_7
   g) ClO
   h) NI_3
   i) SO_3

3. Name each binary acid
   a) HF(aq)
   b) HcL(aq)
   c) HBr(aq)
   d) HI (aq)
   e) H_2S(aq)
   f) H_2Se(aq)
   g) HCN(aq)

4. Name each ternary ionic compound
   a) $Mg(NO_2)_2$
   b) $Al_2(SO_4)_3$
   c) $Ca(OH)_2$
   d) $NaHCO_3$
   e) $KMnO_4$
   f) $Ba(ClO_4)_2$
   g) NH_4NO_3
   h) KH_2PO_4
   i) $Al(BrO)_3$

5. Name each ternary acid
   a) HNO_3(aq)
   b) HNO_2(aq)
   c) H_2SO_4(aq)
   d) H_2SO_3(aq)
   e) $HClO_2$(aq)
   f) $HClO_4$(aq)
   g) H_2CO_3(aq)
   h) H_3PO_4(aq)
   i) HBrO(aq)

6. Give formulas for each binary compound
   a) potassium sulfide
   b) aluminum oxide
   c) nickel(II) chloride
   d) mercury(II) nitride
   e) dinitrogen pentoxide
   f) sulfur dioxide
   g) sulfur hexafluoride
   h) tetraphosphorus decoxide
   i) hydrochloric acid
   j) hydroiodic acid
   k) hydrosulfuric acid
   l) hydroselenic acid

7. Give formulas for each ternary compound
   a) sodium sulfite
   b) calcium carbonate
   c) potassium hydrogenphosphate
   d) aluminum phosphate
   e) iron(II) permanganate
   f) copper(I) hypoiodite
   g) sulfurous acid
   h) carbonic acid
   i) perchloric acid
   j) phosphoric acid
   k) hypofluorous acid
   l) bromic acid

Answers to Review Problems

1.
   a) sodium bromide
   b) aluminum sulfide
   c) potassium oxide
   d) calcium sulfide
   e) magnesium iodide
   f) copper(II) chloride
   g) iron(III) oxide
   h) lead(IV) chloride
   i) mercury(I) chloride

2.
   a) carbon monoxide
   b) carbon dioxide
   c) dinitrogen monoxide
   d) dinitrogen tetroxide
   e) phosphorus pentachloride
   f) dichlorine heptoxide
   g) chlorine monoxide
   h) nitrogen triiodide
   i) sulfur trioxide

3.
   a) hydrofluoric acid
   b) hydrochloric acid
   c) hydrobromic acid
   d) hydroiodic acid
   e) hydrosulfuric acid
   f) hydroselenic acid
   g) hydrocyanic acid

4.
   a) magnesium nitrite
   b) aluminum sulfate
   c) calcium hydroxide
   d) sodium hydrogencarbonate
   e) potassium permanganate
   f) barium perchlorate
   g) ammonium nitrate
   h) potassium dihydrogenphosphate
   i) aluminum hypobromite

5.
   a) nitric acid
   b) nitrous acid
   c) sulfuric acid
   d) sulfurous acid
   e) chlorous acid
   f) perchloric acid
   g) carbonic acid
   h) phosphoric acid
   i) hypobromous acid

6.
   a) K_2S
   b) Al_2O_3
   c) $NiCl_2$
   d) Hg_3N_2
   e) N_2O_5
   f) SO_2
   g) SF_6
   h) P_4O_{10}
   i) $HCl(aq)$
   j) $KI(aq)$
   k) $H_2S(aq)$
   l) $H_2Se(aq)$

7.
   a) Na_2SO_3
   b) CaC_3
   c) K_2HPO_4
   d) $AlPO_4$
   e) $Fe(MnO_4)_2$
   f) $CuIO$
   g) $H_2SO_3(aq)$
   h) $H_2CO_3(aq)$
   i) $HClO_4(aq)$
   j) $H_3PO_4(aq)$
   k) $HFO(aq)$
   l) $HBrO_3(aq)$

Addendum Two
Measurement and Significant Figures

I. Basic Measurements

A. One of the basic tasks of the chemist is to make measurements on the physical and chemical properties of matter

1. Characteristics of measurements

2. Fundamental quantities

3. Examples of fundamental quantities

4. Systems of units

	Quantity	SI unit	Common unit
mass			
length			
time			
temperature			
energy			

5. Greek prefixes can be paired with most of the base units to adjust the size of the unit. Greek prefixes are not used with temperature units.

	Common Greek Prefixes	Abbreviation	Mathematical Equivalent
Mega			
kilo			
deci			
centi			
milli			
micro			
nano			

INTRODUCTION TO CHEMISTRY

Addendum Two

Examples:

1 kg = 1000 g   1 cm = 10^{-2} m

or   or

1 g = 10^{-3} kg   1 m = 10^2 cm

Rule - The positive exponent goes with the smaller unit and the negative exponent goes with the larger unit

II. Scientific Notation

A. Very large and very small numbers can be more conveniently expressed in scientific notation

1. Format for scientific notation

2. The front number

3. The exponent

Examples

$230,000 = 230000 = 2.3 \times 10^{+5}$

$.00046 = .00046 = 4.6 \times 10^{-4}$

III. Calculations using numbers in Scientific Notation

A. Mathematical properties of exponents
Multiplication and Division rule
Exponentiation rule
Addition and Subtraction rule

1. Examples

$10^3 \times 10^4 = 10^{3+4} = 10^7$

$(10^2)^3 = 10^{2 \times 3} = 10^6$

INTRODUCTION TO CHEMISTRY

2. Multiplication and Division of numbers in scientific notation

Examples—

$$(2\times10^3)(4\times10^5) = (2\times4)\times10^8 = 8\times10^8$$

$$\frac{9\times10^{-2}}{3\times10^{-4}} = \left(\frac{9}{3}\right)\times10^{-2-(-4)} = 3\times10^2$$

3. Addition and Subtraction of numbers in scientific notation

Examples—

$$\begin{array}{l} 2\times10^3 \rightarrow 2000 \\ +\ 3\times10^2 \rightarrow 300 \\ \hline 2300 = 2.3\times10^3 \end{array}$$

4. Raising a number to a power

Examples—

$$(3\times10^4)^2 = 3^2\times10^{4\times2}$$

$$= 9\times10^8$$

5. Most of the time putting a number into scientific notation is only a convenience. The only time you must put a number into scientific notation is when it is a whole number ending in a zero.

Examples—

$$5200 = 5.200\times10^4$$
$$= 5.20\times10^4$$
$$= 5.2\times10^4$$

The number of zeros kept depends on the uncertainty in the measurement

Addendum Two

IV. Uncertainty in Measurements

    A. All measurements involve some degree of uncertainty.

    B. The manner in which the measuring device is graduated determines the number of reported digits

    C. Examples

    D. Significant digits

    E. Two terms relating to uncertainty which are often confused are accuracy and precision

                      accuracy vs. precision

    F. Examples

G. Errors existing in an experimental procedure may be either systematic or random

Systematic error vs. Random error

1. A reported measured value must be thought of as a range

Examples—

$$12.6 \text{ mL} = (12.6 \pm 0.1 \text{ mL})$$
$$= (12.5 - 12.7 \text{ mL})$$

It is important to be able to determine the number of significant digits in a given reported measurement

V. Rules for determining significant digits in a measurement

1. Non-zero digits are always significant

Examples: 13.75 m = 4 sig. figs.

2. Zero digits may or may not be significant

   a) Captive zeros are significant

   Examples : 5.03 g = 3 sig. figs.

   b) Trailing zeroes in a decimal point number are significant

   Examples: 2.40 cm = 3 sig. figs.

   c) Leading zeros are not significant

   Example: 0.004 s = 1 sig. fig.

   d) Trailing zeros in a whole number may or may not be significant
   These values must be put into scientific notation

   Examples: 240 g = 2 or 3 sig. figs.

Addendum Two

3. Exact or defined numbers have infinite significant figures

   Examples: 15 beakers = infinite sig. figs.

   The number of significant digits or decimal places which can be kept in the final answer of a calculation depends on the type of calculation and the number of significant digits or decimal places in the numbers involved in the measurement.
   In order to apply the rules we are about to discuss you need to be able to round-off properly

VI. Rounding-off Rules

1. Rounding off from a number less than 5

   Examples: 5.34 to 2 sig. figs.
   5.3 ¦ 4 = 5.3

2. Rounding off from a number greater than 5

   Examples: 62.47 to 3 sig. figs.
   62.4 ¦ 7 = 62.5

3. Rounding off from a 5 followed by one or more non-zero digits

   Examples: 13.52 to 2 sig. figs.
   13. ¦ 52 = 14

4. Rounding off from an exact 5 — Even or odd rule
   Examples: 29.245 to 4 sig. figs.
   29.24 ¦ 5 = 29.24

5. Sometimes rules 3 and 4 are combined into rule 2

VII. Significant digit rules for multiplication and division

    A.   **Examples**

$$(2.0)(4.136) = 8.2|72 = 8.3 \text{ to 2 sig. figs.}$$

2 sig. figs. 4 sig. figs.

VIII. Significant digit rules for addition and subtraction

    A.   Examples

14.135	3 places past decimal
8.2	1 place past decimal
22.335	

= 22.3 to 1 place past decimal

Things to Know

Definitions:

1. Measurement
2. Fundamental quantity
3. Scientific notation
4. Significant digits
5. Accuracy
6. Precision
7. Systematic error
8. Random error

Concepts

1. Give examples of fundamental quantities
2. Give SI and common unit for mass
3. Give SI and common unit for length
4. Give SI and common unit for time
5. Give SI and common unit for temperature
6. Give SI and common unit for energy
7. Give abbreviation and math equivalent for Mega, kilo, deci, centi, milli, micro, nano
8. Be able to put a number into scientific notation
9. Be able to take a number in scientific notation and expand it into normal form
10. Be able to multiply and divide numbers in scientific notation
11. Be able to add and subtract numbers in scientific notation
12. Be able to raise a number in scientific notation to a power
13. Understand how the manner in which a measuring device is graduated determines the number of reported digits
14. Understand how to interpret a reported measured value as a range
15. Know the significant figures rules for non-zero and zero digits
16. Be able to apply the significant figure rules to any given measured value
17. Understand the difference between measured and exact numbers
18. Know the significant figure rule for multiplication and division. Be able to apply it
19. Know the significant figure rule for addition and subtraction. Be able to apply it
20. Know the round-off rules. Be able to apply them

Review Questions

1. Measurements have a _____ part and a _____ part
2. A measured quantity which is independent of others is a _____ quantity
3. Three examples of fundamental quantities are _____, _____, _____
4. The two main systems of units are the _____ units and the _____ units
5. For mass the SI unit is the _____ and the common unit is the _____
6. For length the SI unit is the _____ and the common unit is the _____
7. For time the SI unit is the _____ and the common unit is the _____
8. For temperature the SI unit is the _____ and the common unit is the _____
9. For energy the SI unit is the _____ and the common unit is the _____
10. Very large and very small numbers are more conviently expressed in _____
11. The front number in scientific notation format must lie between _____
12. The exponent in scientific notation format must be a _____ or _____ integer
13. The only time you must put a number into scientific notation is a _____ number ending in a _____
14. All measurements involve some degree of _____
15. The number of digits which can be reported in a measurement depends on how the measuring device is _____
16. All digits measured exactly plus one estimated digit are called _____ digits
17. How close a measurement is to the true value is called the _____
18. How close a series of measurements are to each other is called the _____
19. Errors which can be identified and eliminated are called _____ errors
20. Errors which are inherent and can't be eliminated are called _____ errors
21. Systematic error affects the _____ of the result
22. _____ error affects the precision of the result
23. All _____ digits are significant
24. Captive zeros are _____
25. _____ zeros are not significant
26. Trailing zeros in a decimal point number are _____
27. Trailing zeros in a whole number _____ or _____ be significant
28. _____ numbers have infinite significant figures
29. Exact numbers can result from _____ or from _____

Addendum Two

Review Problems

1. For each use the mathematical value of the Greek prefix to determine the relationship.
   a) 1 km = ? m
   b) 1 cs = ? s
   c) 1 mg = ? g
   d) 1 g = ? μg
   e) 1 J = ? kJ
   f) 1 s = ? ns
   g) 1 m = ? cm
   h) 1 ng = ? g
   i) 1 MJ = ? J

2. Convert each of the following numbers into scientific notation
   a) 134
   b) 0.358
   c) 678,000,000
   d) 0.03640
   e) 0.0000005
   f) 18,000
   g) 35,000,000,000,000
   h) 36×10^4
   i) 175×10^{-5}

3. Perform each of the following calculations
   a) $(2 \times 10^4)(4 \times 10^{-2})$
   b) $(1.5 \times 10^{-3})(6 \times 10^7)$
   c) $(8 \times 10^{-1})(5 \times 10^3)$
   d) $\dfrac{6.0 \times 10^4}{3.0 \times 10^5}$
   e) $\dfrac{1.2 \times 10^2}{4.0 \times 10^1}$
   f) $(2.0 \times 10^3)^{+4}$
   g) $6.0 \times 10^3 + 2.0 \times 10^2$
   h) $5.0 \times 10^5 + 3.0 \times 10^5$
   i) $1.2 \times 10^4 + 6.0 \times 10^5$

4. Determine the number of significant digits in each of the following
   a) 126
   b) 404
   c) 0.027
   d) 0.370
   e) 304.00
   f) 0.0006
   g) 4.2×10^4
   h) 21 eggs
   i) 12 inches = one foot

5. Round-off each number to the indicated number of significant figures
   a) 348.3 (3)
   b) 141.7 (3)
   c) 0.0368 (2)
   d) 221.5 (3)
   e) 64.5 (2)
   f) 63.5 (2)
   g) 122.501 (3)
   h) 142 (2)
   i) 337 (1)

6. Perform each indicated calculation and round-off to the proper number of significant figures (assume all values are measured)
   a) 123×4.0
   b) 2.0365×12.04
   c) 5.0×118.03
   d) $\dfrac{124.3}{3.0}$
   e) $\dfrac{5.4 \times 10^4}{2.35 \times 10^3}$
   f) $\dfrac{9.0 \times 10^4}{2.375 \times 10^3}$
   g) $2.04 + 8.3175$
   h) $18.4256 - 0.021$
   i) $23.419 - 6$

INTRODUCTION TO CHEMISTRY

7. Interpret each reported measured value as a range
   a) 3.4 mL
   b) 8.14 cm
   c) 12 L
   d) 5.68 g
   e) 32 s
   f) 12.0 kJ

Answers to Review Problems

1.
- a) 1000
- b) 10^{-2}
- c) 0.001
- d) 10^6
- e) 10^{-3}
- f) 10^9
- g) 100
- h) 10^{-9}
- i) 10^6

2.
- a) 1.34×10^2
- b) 3.58×10^{-1}
- c) 6.78×10^8
- d) 3.64×10^{-2}
- e) 5×10^{-7}
- f) 1.8×10^4
- g) 3.5×10^{13}
- h) 3.6×10^5
- i) 1.75×10^{-3}

3.
- a) 8×10^2
- b) 9×10^4
- i) 4.0×10^3
- d) 2.0×10^{-1}
- e) 3.0×10^0
- f) 1.6×10^{13}
- g) 6.2×10^3
- h) 8.0×10^5
- i) 6.12×10^5

4.
- a) 3
- b) 3
- c) 2
- d) 3
- e) 5
- f) 1
- g) 2
- h) ∞
- i) ∞

5.
- a) 348
- b) 142
- c) 0.037
- d) 222
- e) 64
- f) 64
- g) 123
- h) 1.4×10^2
- i) 3×10^2

6.
- a) 4.9×10^2
- b) 24.52
- c) 5.9×10^2
- d) 41
- e) 23
- f) 3.8×10^1
- g) 10.36
- h) 18.405
- i) 17

7.
- a) (3.3—3.5) mL
- b) (8.13—8.15) cm
- c) (11—13) L
- d) (5.67—5.69) g
- e) (31—33) s
- f) (11.9—12.1) kJ

Addendum Three
Problem Solving Techniques

Most problems in chemistry can be solved either by dimensional analysis or application of a formula

I. Dimensional analysis

    A. Set up problem in proper format

    B. Set up unit factor first in terms of units

    C. Determine mathematical relation between numerator and denominator units

II. Unit Conversions

    A. Sources of information for proper mathematical relation

III. One step unit conversions

Examples:

Convert 13 cm to m

$$13 \text{ cm} \left(\frac{1 \text{ m}}{10^2 \text{ cm}} \right) = 0.13 \text{ m}$$

Convert 4.6×10^{-4} g to µg

$$4.6 \times 10^{-4} \text{ g} \left(\frac{10^6 \text{ µg}}{1 \text{ g}} \right) = 4.6 \times 10^2 \text{ µg}$$

More examples:

IV. Two step unit conversions

Examples:

Convert 3.1×10^{-5} km to mm

$$3.1 \times 10^{-5} \text{km} \left(\frac{10^3 \text{m}}{1 \text{km}}\right)\left(\frac{10^3 \text{mm}}{1 \text{m}}\right) = 31 \text{ mm}$$

Convert 9.3×10^3 nm to cm

$$9.3 \times 10^3 \text{nm} \left(\frac{1 \text{m}}{10^9 \text{nm}}\right)\left(\frac{10^2 \text{cm}}{1 \text{m}}\right) = 9.3 \times 10^{-4} \text{ cm}$$

More examples:

V. Unit conversions involving squares, cubes or higher powers and ratios of units

Examples:

Convert $4.7 \times 10^2 \text{ cm}^2$ to m^2

$$47 \times 10^2 \text{ cm}^2 \left(\frac{1 \text{m}}{10^2 \text{cm}}\right)^2 = 4.7 \times 10^2 \text{ cm}^2 \left(\frac{1 \text{m}^2}{10^4 \text{cm}^2}\right) = 4.7 \times 10^{-2} \text{ m}^2$$

More examples:

VI. Density

A. Density is a derived unit as is volume

Derived unit	SI unit	Common unit
volume		
density		

VII. Density problems—Given mass and volume

Example:

An object has mass = 20.0g, volume = 4.0 mL. Calculate the density

$$d = \frac{m}{V} = \frac{20.0\text{g}}{4.0\text{mL}} = 5.0 \text{ g/mL}$$

More examples:

Addendum Three

VIII. Density problems—Given volume and density

Example: An object has a volume of 20.0mL and density of 4.00g/mL

By algebraic rearrangement: $d = \dfrac{m}{V} \rightarrow V \cdot d = \dfrac{m}{V} \cdot V \rightarrow m = d \cdot V$

$m = (4.00\text{g/mL})(20.0\text{mL}) = 80.0\text{g}$

By dimensional analysis: Restate problem as 20.0mL = ? g

$20.0\text{mL}\left(\dfrac{4.00\text{g}}{1\text{mL}}\right) = 80.0\text{g}$

More examples:

IX. Density problems—Given mass and density

Example: An object has mass = 10.0g and density = 2.0g/mL. Determine its volume.

By algebraic rearrangement:

$d = \dfrac{m}{V} \rightarrow V \cdot d = \dfrac{m}{V} \cdot V \rightarrow \dfrac{V \cdot d}{d} = \dfrac{m}{d} \rightarrow V = \dfrac{m}{d}$

$V = \dfrac{10.0\text{g}}{2.0\text{g/mL}} = 5.0\text{mL}$

By dimensional analysis: Restate problem as 10.0g = ? mL

$10.0\text{g}\left(\dfrac{1\text{mL}}{2.0\text{g}}\right) = 5.0\text{mL}$

Addendum Three

More examples:

Things to Know

Concepts
1. Know the general approach of dimensional analysis
2. Be able to do one step unit conversions
3. Be able to do two step unit conversions
4. Be able to do unit conversions involving squares and cubes
5. Know the density formula
6. Know the SI and common units for volume
7. Know the SI and common units for density
8. Given mass and volume be able to calculate density
9. Given density and volume be able to calculate mass by algebraic rearrangement
10. Given density and volume be able to calculate mass by dimensional analysis
11. Given density and mass be able to calculate volume by algebraic rearrangement
12. Given density and mass be able to calculate volume by dimensional analysis

Review Questions

1. What is the formula for density _____
2. What is the SI unit for volume _____
3. What is the SI unit for density _____
4. The common unit for volume is _____
5. The common unit for density is _____

Review Problems

1. Perform each of the following conversions
   a) $3.0 \text{ g} = ? \text{ kg}$
   b) $27.3 \text{ L} = ? \text{ mL}$
   c) $2.0 \times 10^4 \text{ m} = ? \text{ nm}$
   d) $3.0 \times 10^{-4} \text{ km} = ? \text{ m}$
   e) $5.0 \times 10^{-2} \text{ mg} = ? \text{ g}$
   f) $3.7 \times 10^3 \text{ μL} = ? \text{ L}$

2. Perform each of the following conversions
   a) $5.3 \times 10^3 \text{ μm} = ? \text{ mm}$
   b) $6.4 \times 10^{-2} \text{ cg} = ? \text{ ng}$
   c) $1.2 \times 10^5 \text{ ns} = ? \text{ ks}$
   d) $6.4 \times 10^{-7} \text{ kg} = ? \text{ μg}$
   e) $9.9 \times 10^3 \text{ MJ} = ? \text{ dJ}$
   f) $7.1 \times 10^{-4} \text{ mm} = ? \text{ cm}$

3. Perform each of the following conversions
   a) $2.0 \text{ m}^2 = ? \text{ cm}^2$
   b) $3.0 \times 10^3 \text{ mL}^3 = ? \text{ L}^3$
   c) $5.0 \times 10^8 \text{ cm}^3 = ? \text{ m}^3$

4. Determine the density if
   a) mass = 24.0g, volume = 8.0mL
   b) mass = 3.00g, volume = 10.0mL
   c) mass = 8.00g, volume = 2.000mL

5. Determine the mass if
   a) volume = 2.00mL, density = 4.0g/mL
   b) volume = 24.0mL, density = 1.50g/mL
   c) volume = 5.0mL, density = 9.00g/mL

6. Determine the volume if
   a) mass = 12.0g, density = 1.2g/mL
   b) mass = 4.00g, density = 8.00g/mL
   c) mass = 2.00×10^2 g, density = 5.0g/mL

Answers to Review Problems

1.
- a) 3.0×10^{-3} kg
- b) 2.73×10^{4} L
- c) 2.0×10^{13} nm
- d) 3.0×10^{-1} m
- e) 5.0×10^{-5} g
- f) 3.7×10^{-3} L

2.
- a) 5.3 mm
- b) 6.4×10^{5} ng
- c) 1.2×10^{-7} ks
- d) 6.4×10^{2} µg
- e) 9.9×10^{10} dJ
- f) 7.1×10^{-5} cm

3.
- a) 2.0×10^{4} cm^2
- b) 3.0×10^{-6} L^3
- c) 5.0×10^{2} m^3

4.
- a) 3.0 g/mL
- b) 0.300 g/mL
- c) 4.00 g/mL

5.
- a) 8.0 g
- b) 36.0 g
- c) 45 g

6.
- a) 1.0×10^{1} mL
- b) 0.500 mL
- c) 4.0×10^{1} mL

Addendum Three